Praise for Lisa Steele and *Duck Eggs Daily*

"As someone who has kept ducks off and on my entire life I find it bewildering that they are often overlooked as candidates for the home flock. They can be easy to keep, prolific egg layers and endlessly amusing to have around. Lisa and I share equally our enthusiasm for ducks. Lisa's book sets the homeowner on the right path, providing both the knowledge and confidence for success."

~ P. Allen Smith, host of PBS's "Garden Home"

"When it comes to ducks, Lisa Steele knows her stuff, offering up the insight and advice you will need to raise happy, healthy ducks. Whether you're nurturing ducklings or wondering what to do with your windfall of duck eggs, Steele's book offers gentle guidance on the art and science of keeping these beautiful, rewarding animals."

~ Felicia Feaster, Editor-in-Chief, HGTV Gardens

"If you're ready to move from the wonderful world of backyard chicken-keeping to raising ducks, you couldn't ask for a better friend and mentor to have by your side than Lisa Steele. With the care and attention she gives every member of her flock and the sage advice she passes along to the poultry community, any bird in her care is, well, a lucky duck."

~ Rachael Brugger, Managing Editor, HobbyFarms.com and UrbanFarmOnline.com

"It's no surprise that Lisa should expand her experience of backyard chickens to keeping ducks and once again she's providing us with sound advice based on her no-nonsense, firsthand experience. Great resource if you favor a natural approach to raising ducks."

~ Pascale Deffieux-Pearce, Executive Vice-President, Brinsea Products, Inc.

"I have a duck and goose farm and sell hundreds of thousands of ducklings and goslings a year – yet I found Lisa's book fascinating and couldn't put it down! Great description of ducks and their lives, evident hands-on knowledge, fantastic suggestions, very practical recommendations, and charming observations of her ducks and their actions. I highly recommend this book to anyone who is considering a quack, a splash, and an entertainer in their back yard or on their farm."

~ John Metzer, owner, Metzer Farms

"For all you duck lovers and duck farmers, here is your next "must read" book. It's filled with great resources and all the information you need to know about selecting and caring for your flock. Lisa Steele has been rocking the chicken world for years at Fresh Eggs Daily and now she is sharing her passion and wealth of knowledge about ducks with you."

~ C S Wurzberger, The Green Up Girl, podcast host
for The Livestock Conservancy's "Heritage Breeds"

DUCK EGGS
Daily

DUCK EGGS
Daily

raising happy, healthy ducks... **naturally**

LISA STEELE

st. Lynn's press

PITTSBURGH

Duck Eggs Daily
Raising Happy, Healthy Ducks…Naturally

ISBN-13: 978-0-9892688-8-2

Library of Congress Control Number: 2015938236
CIP information available upon request

First Edition, 2015

St. Lynn's Press . POB 18680 . Pittsburgh, PA 15236
412.466.0790 . www.stlynnspress.com

Book design – Holly Rosborough
Editor – Catherine Dees

Photo credits: pg. ix and 45 © Lee Shilling Edwards; pg.27 © Erica Deeds Kellam;
pgs. 31, 39 and 113 © Sarah Barrera, Chicken Boots LLC;
pgs. 56, 134 and 135 ©Tracy at Duck Duck Dog; pg. 77 © David Allen Cooper;
pg. 81 © Ann, A Farm Girl in the Making; pg. 90 © Chloe's Creek; pg. 134 © Amy Fewell,
The Fewell Homestead; pg. 135 © Tonya Mappin – Mappin's Waterfowl;
pgs. 134 and 135 © Darlene Terry, Whimsical Years Photography
All other photography © Lisa Steele

Printed in China
on certified FSC recycled paper using soy-based inks

This title and all of St. Lynn's Press books may be purchased for educational,
business or sales promotional use. For information please write:
Special Markets Department . St. Lynn's Press . POB 18680 . Pittsburgh, PA 15236

10 9 8 7 6 5 4

For
Puddles
and
Bob

TABLE OF CONTENTS

INTRODUCTION

Raising ducks wasn't part of the original plan when I walked into the feed store that fateful morning early in the spring of 2009 to pick out some day-old chicks to begin my foray into backyard chicken keeping. But when we finished picking out our selection of chicks and my husband said, "We'll take two of those, too" and pointed towards a bin full of peeping ducklings, my heart melted; I couldn't resist – especially after the feed store owner told us that raising ducks is exactly the same as raising chickens (which turned out to not be entirely true, but more on that later!). I had grown up around chickens and raised them as a child, so I felt pretty comfortable I could handle them; ducklings were another story. I had no experience raising ducks. Figuring it couldn't be that hard, we put two of the fuzzy brownish ducklings into the box with our chicks and happily made our way home.

I spent the next few weeks scouring the Internet for information on raising ducks and read every book and magazine article I could get my hands on. As it turned out, for optimal duck health and happiness, there are a few things that don't come into play in chicken keeping…

The first thing that became painfully obvious was that ducklings grow fast – really fast. Within a few days, the tiny ducklings that had started out the same size as the chicks when we bought them were towering over the poor chicks and trampling them. Not only that, the ducklings would immediately drain the chick waterer every time I filled it and the brooder was constantly a mess of sodden shavings. I switched to a shallow water bowl and the ducklings would sit in it and splash to their hearts' content, not allowing the chicks near to get a drink. And that was the end of community brooding. The ducklings moved into a spare bathtub until they were ready to be outside, leaving the chicks alone – and finally dry and happy.

Despite that slightly rough start, I have enjoyed every minute of raising ducks. Those two Mallard ducklings, Puddles and Bob, were soon joined by three Pekin ducklings from a local farm that were supposed to be an Easter surprise for a buyer who never showed up to claim them. I started researching different domestic duck breeds and realized I would have to either order shipped ducklings or buy some fertile hatching eggs to get the breeds I wanted, since only Pekins and Mallards were available to me locally. That led to more reading and research into hatching duck eggs and soon I had several successful duckling hatches under my belt using an incubator. I even got lucky last spring and hatched eggs under a duck – my first broody duck ever!

Over the years, I have learned a lot about brooding ducklings and raising ducks through constant reading and my own trial and error. I use all-natural methods in raising our ducks, incorporating lots of weeds, herbs and natural supplements to help build strong immune systems and produce healthy eggs. Our grown ducks are raised alongside our chickens and they all co-exist peaceably, although the ducks now sleep in separate quarters for reasons that will be covered in this book. The ducks add so much personality and enjoyment to our little barnyard that I can't imagine our chicken run without them.

I decided to write this book to share my experiences with anyone who is interested in raising ducks. As you have already gathered, whether or not you already raise chickens, there is a bit of a learning curve. Yes, the basics are the same, but I'll show you some easy things that you can do to ensure the happiest, healthiest ducks possible! Backyard ducks aren't nearly as popular as backyard chickens – yet – but hopefully after you read this book, you'll fall in love with ducks as I have and decide to raise a few in your backyard, too.

Lisa

ADDING DUCKS TO A CHICKEN FLOCK

Do you already raise chickens and are thinking about adding a few ducks to your flock? It's as easy as, well, adding a few ducks. I've introduced ducklings (at least 8-10 weeks old so they're big enough to not be trampled or drowned accidentally) and full-grown ducks to my mixed chicken and duck flock – which includes both a rooster and a drake, by the way – and never had any problems.

I think the chickens instinctively know that ducks are no threat to their pecking order. My chickens have never seemed concerned about sharing space with ducks. As for the ducks, they don't adhere to nearly as strict a pecking order as do chickens, and they seem

to view new ducks as more pool party participants; there's always room for a few more!

But... Always separate new ducks for a few weeks to be sure they are not sick and won't transmit anything to your existing flock, and to let them get used to their new home slowly; after that, you shouldn't have any issues when you introduce them. And always watch for a while when you first add your ducks to be completely sure there won't be any pecking or feather pulling or bullying. If you work during the week, first thing Saturday morning would be a good time to let your ducks get to know the chickens and vice-versa.

Ducks are very welcoming in general to newcomers. The exception could be a new drake introduced to a small flock

with an existing drake, as some fighting over females could ensue, so always be sure to keep a close eye any time you add new flock members. Ducklings should never be added to a flock of grown ducks (unless accompanied by their mother duck to protect them) until they are nearly full grown, to prevent drowning or trampling accidents.

Sharing the coop. As for sleeping quarters, your ducks will be perfectly happy bunking in the chicken coop on the floor in the straw. They will even make a nest of their own in a corner in which to lay their eggs. Ducks really are low-maintenance, so if you've already got a coop with chickens in it, you don't need to do a thing to accommodate a few ducks (but see page 75 for some differences to be aware of).

Feed and Treats. Our chickens and ducks eat the same feed and I give them the same treats. (For their particular preferences, see page 61.)

Feeders and waterers. You may need to rethink your feeders and waterers. (For the why and how, see pages 34 and 35.)

Adding a pool. You will also want to set up a pool area in a far corner of your run. A kiddie pool works well since it's deep enough for the ducks to enjoy, but not so deep that a chicken will drown if she falls in. My chickens learned early on to avoid the mud around the pool area, but I do occasionally find a hen perched on the side of the pool.

Besides that, your chicken flock will greatly benefit from the addition of a few ducks, with very few changes made to your existing set-up.

❦ *Optimal Duck Flock Size* ❦

Single Duck – Not recommended, since they are social animals.

A Pair – Okay, but not optimal. If you lose one, the other will be inconsolable. Male/female pair can result in over-mating; two females is a better pair choice if two ducks is all you can have.

A Trio – One drake and 2 ducks can also lead to over-mating but is a good "starter" flock size and usually works well, as does 3 females.

Half Dozen – One drake and 4-5 ducks: my favorite flock size. Enough ducks to prevent over-mating; lots of eggs.

Larger flock – Multiple drakes with a minimum of 4-5 ducks per drake. If you keep multiple drakes, be sure you have enough ducks to give each his own mini-flock, although this can still lead to fighting or over-mating of a "favorite" female.

All Females (any number) – Can be noisy, but for egg production this is the optimal flock, although the eggs won't be fertile.

All Drakes – Great if you want the ducks primarily as pets and don't care about eggs. They are still great for bug control and hours of enjoyment. Also, since the drakes are very quiet, an "all boy" flock is good if you have close neighbors. Drakes generally won't fight if there are no females present, and drakes are often easy to find for free at shelters, on Craigslist or at local farms.

ONE

WHAT'S SO SPECIAL ABOUT DUCKS?

There is growing interest in raising backyard ducks, whether as an addition to a flock of chickens or on their own. I can vouch for the fact that ducks are a wonderful addition to any back yard, having raised both chickens and ducks for several years now side-by-side, co-mingling them by day in a large run and providing separate sleeping quarters. Although they are becoming more popular, backyard ducks still are running a distant second to backyard chickens. But more and more people seem to be considering adding ducks into their lives.

Ducks, like chickens, come in a multitude of shapes, sizes and colors. If you are thinking of starting a backyard flock, you might want to consider some of the outstanding breeds that are on the Livestock Conservancy Critical List (meaning that fewer than 500 ducks of breeding stock exist in the United States, with five or fewer breeding flocks of 50 ducks or more, and fewer than 1,000 ducks worldwide), including Saxonies, Anconas, Magpies and Silver Appleyards – all of which I raise. Every backyard flock that starts up and includes some of these endangered breeds helps to ensure that the breed will continue to flourish. Raising the awareness of many of these rare breeds has been one of my goals with my own flock and one of my reasons for writing this book.

Over the years, I have had ample opportunity to observe both our chickens and ducks, and I think there are some distinct advantages to raising ducks, especially for the small backyard enthusiast. Here are some reasons why I actually prefer raising ducks to raising chickens:

Ducks are quieter

Ducks are much quieter than chickens - even in their run.

One benefit of raising ducks is that they don't feel the need to loudly proclaim to the world every time they lay an egg, like a chicken does; instead, they stealthily lay their eggs pre-dawn under cover of darkness, in complete silence. This is especially important for those who live in neighborhoods with other houses close by. Chickens cackle and carry on after they lay an egg, before they lay an egg, and for no apparent reason at all. Female ducks, on the other hand, although they will quack loudly when agitated or excited, normally just quietly chitter-chatter among themselves. And unlike roosters – who seem to feel compelled to communicate their presence, not just at dawn, but throughout the day – drakes (male ducks) don't have a real quack. They make only a soft raspy noise.

Ducks are generally healthier overall

Ducks have hardier immune systems, tend to stay in better overall health and are less likely to contract diseases than chickens. Ducklings generally don't contract either coccidiosis or Marek's, two serious concerns for baby chicks, and adult ducks aren't as susceptible to external parasites because they spend so much of their time in the water – so mites, lice and other external parasites that might be tempted to latch on will drown. Ducks run an extraordinarily high normal body temperature of around 107°F, which makes their body inhospitable to most pathogens and resistant to disease.

Ducks are more heat-tolerant

During the hot and humid summer months when I lived in Virginia, our chickens would stand around panting, trying to stay cool in the shade. Ducks handle the heat quite easily by merely taking a dip in their pool to cool off. They paddle about contentedly, hopping out only to enjoy some chilled watermelon or other summertime treats.

Ducks are more cold-hardy

Ducks have waterproofing on their feathers to protect against the elements, as well as a thick down undercoat designed to keep them warm and dry in the water and rain. This makes ducks far more cold-hardy than chickens. In fact, our ducks actually prefer to sleep outside, even in the snow and inclement weather. Ducks also have an added layer of fat that chickens don't have.

Duck eggs are superior to chicken eggs

Ducks lay eggs that are larger, richer in flavor and excellent for baking, due to their higher fat and lower water content. And duck eggs are slightly more nutritious than chicken eggs. Pastry chefs prize duck eggs because the large amount of protein in their whites adds heft and loft to baked goods. With their thicker shells and membranes, they also stay fresh longer and are less likely to break.

Ducks don't mind being out in the cold.

Ducks lay more regularly

You get more eggs with ducks than chickens.

Our ducks consistently outperform our chickens in egg production, even through the winter without any supplemental light in their house. Most domestic ducks are also very unlikely to go broody (broodies stop laying eggs and therefore become unproductive, unless you're trying to hatch eggs). Ducks generally lay productively into their fourth year and continue to lay for several years after that.

Ducks adhere to a far less aggressive pecking order

Drakes are not nearly as aggressive as roosters and rarely turn nasty toward humans. Ducks also don't take their pecking order as seriously as chickens and tend to welcome newcomers far more quickly – and with far less squabbling than do chickens. Whether the newcomers are chickens or ducks, our ducks seem unperturbed by it all and seldom bother new additions to the flock. Ducks won't bicker within the flock about pecking order or other issues. They are far more laid back in temperament.

Ducks are easier on your lawn

Ducks don't scratch grass or plants down to the bare dirt like chickens do. They may trample your lawn a bit and will dabble in the mud around their water tubs or in the dirt looking for snacks, creating small, deep holes in the ground, but they won't turn your backyard into a barren wasteland like chickens will.

It's true that ducks can, and will, eat anything green within their reach, but…as long as you plant bushes and trees that are tall enough so that the ducks can't reach the tops – or fence around your gardens and flowering plants – you can successfully landscape your run or backyard, even with ducks inhabiting it. Fencing around your gardens and landscaping doesn't need to be much more than two or three feet high because most domestic duck breeds can't fly, and ducks generally can't hop or flutter and flop as high as chickens can.

Ducks are wonderful for pest control

Ducks will eat every slug, grub, earthworm, spider, grasshopper, cricket, and beetle they can find in your yard. Given the opportunity, they will also eat snakes, mice, frogs, and lizards. Ducks are wonderful for natural pest control, but be aware they will eat so-called "good" bugs and beneficial worms and toads as well.

Ducklings are adorable!

Okay, I admit this comes down to personal preference. As cute as baby chicks are, baby ducklings are irresistible. Those too-big-for-their-bodies webbed feet, earnest dark eyes and almost flesh-colored rounded bills steal my heart every time. As a bonus, ducklings grow up to be adorable ducks!

On the flip side, sure, ducks can be pretty willful and stubborn. They don't automatically put themselves to bed at dusk like chickens do (although they can be trained to head to their house each night without much trouble). They also are messier than chickens when it comes to water and mud, but the ducks themselves are always pristine and perfectly clean, even our snow-white Pekins! I find ducks pretty unlikely to get their feathers ruffled for the most part. They are generally calm, alert, always happy and downright funny.

Are you considering a few ducks yet? You will want to invest in a few pairs of muck boots, most definitely, but it's a small price to pay for the joy a flock of ducks brings. In my eyes, ducks win out as my top choice for a backyard flock and they will always be an important part of ours.

There is no such thing as an ugly duckling.

DUCKS VS. CHICKENS

	CHICKENS	DUCKS
Approx. Date of First Egg	18-24 weeks	21-24 weeks
Productive Laying Life	2-3 years	5-6 years
Average Life Span	8-10 years	10-15 years
Laying Season (*without added light*)	Spring-fall	Year-round
Daily Sunlight Requirements	15-17 hours	14-16 hours
Average Annual Egg Production	250	300+
Egg Color	Varied	Slightly varied
Egg Size	Medium-large	Extra-large
Tendency to Brood	Average	Low
Noise Level:		
Hens/Ducks	Low-medium	Low-Medium
Roosters/Drakes	High	Extremely low
Damage to Lawn/Landscaping	Heavy	Low
Manure for the Garden	"Hot" – needs to age for 6 months	Ready immediately
Heat Tolerance	Fair	Good
Cold Hardiness	Good	Excellent
Inclement Weather Tolerance	Fair	Excellent
Disease Resistance	Fair	Excellent
Parasite Resistance	Fair	Excellent
Predator Risk	High	Extremely high
Minimum Space Requirements		
Coop/House	2-4 sq ft/hen	3-5 sq ft/duck
Run	10 sq ft/hen	15/sq ft/duck
Water Requirements	Average	High

TWO

HATCHING DUCKLINGS

Hatching your own ducklings is an inexpensive, relatively easy way to start or add to your backyard flock. It's an awe-inspiring experience to watch the embryos develop and the ducklings hatch. I prefer hatching my own ducklings in an incubator. Not only is it the easiest way to add rare breeds to your flock that aren't available locally, I believe the ducklings do imprint on me and end up being far friendlier than those I buy as day-old ducklings – or ducklings that have hatched under a broody duck. Domestic ducks rarely go broody (i.e., sit on fertile eggs until they hatch) anyway, so hatching eggs in an incubator is generally your best bet. Various types of incubators work slightly differently, so it's important to read the instruction manual for your particular model. Here are some general tips for a successful hatch.

Obtaining eggs for hatching

If you have fertile eggs from your own ducks, you will likely have the best hatch rate. If not, be sure to order your hatching eggs from a reputable breeder or hatchery. If you can find eggs from a local farm, that's even better. Shipped eggs are often jostled or subjected to temperature and humidity fluctuations and have a far lower hatch rate than eggs that don't have to be shipped. An 80% hatch rate is considered very good; for shipped eggs, the norm is closer to a 50-60% hatch rate.

Most problems with eggs not hatching can be attributed to old eggs with low fertility, rough handling, eggs stored at an improper temperature, improper turning, uneven incubator temperature or humidity, or nutritional deficiencies in the breeding stock.

If you have your own fertile eggs (laid by your own ducks), pick those that are most perfectly and uniformly shaped, preferably not covered with mud or manure. Don't wash them; instead, carefully scrape off any muck with your fingernail or a rough sponge. Don't choose overly small or large eggs because they tend to not hatch well.

Before you turn on the incubator

While you're collecting enough to fill your incubator, store them, pointy end down, at a 45-degree angle in a cool location – around 60°F is optimal. Don't turn on the incubator yet! Rotate the eggs side-to-side several times a day to keep the yolk centered in the white. Hatchability declines each day after an egg is laid. Fertile eggs will stay viable for about a week after being laid. After that, fertility starts to decline, so try not to delay setting the eggs for too long.

Check for cracks in the shell using the "candling" method.

"Candling." "Candle" each egg one by one, whether you are collecting your own or you ordered shipped eggs. Candling is so called because in early days before electricity, the light from a candle flame was used to check for cracks, and later during the incubation period to "see" inside the egg to monitor development. You are checking for hairline cracks. You can use a regular flashlight – no need to use candles any longer! Cup your hand around the beam to shine it through the shell, or buy a commercial egg candler. Discard any cracked egg or seal minor cracks with softened beeswax to prevent bacteria and air from entering the egg through the crack and killing the embryo.

In the incubator

When you are ready to set your eggs, turn your incubator on and let it get up to temperature while you let the eggs sit for several hours at room temperature to allow them to warm up a bit and the yolks to settle. Mark an X or number on one side of each egg with a pencil. This will serve as your guide when turning the eggs (if your incubator has an auto turner, you can skip this step). Place the eggs in the incubator with the pointy end angled down.

Marking the eggs with a pencil serves as a guide for turning them.

Set your incubator in a quiet location out of direct sunlight where it won't be bothered by children or pets. I like to put a piece of rubber shelf liner on the floor of the incubator. This helps the eggs stay in place as you turn them and also gives the newly hatched ducklings a textured surface to grip onto so they don't slip.

Managing temperature and humidity. Duck eggs should be incubated at a temperature between 99.3° and 99.6°F (but again, check the setting for your particular model) for 28 days. The humidity level in the incubator is extremely important as well and needs to be monitored. Depending on the type of incubator you are using, the humidity can be controlled by filling small water reservoirs, or wetting a clean kitchen sponge and setting it inside the incubator. Humidity should be checked using a hygrometer, available from your feed store or online, if your incubator doesn't come equipped with one, and kept constant according to your incubator instruction manual.

Moisture is lost through the pores in the eggshell, and air is drawn in. The air sac in the egg gets larger as an egg ages, whether the egg is being incubated or not. It's crucial the air sac be the correct size in a developing egg to allow the embryo room to grow and air to breathe before it hatches. If the humidity is too high in the incubator, the air sac will be too small and the duckling can be too large and have trouble breathing and breaking out of the shell. Conversely, low humidity will result in a larger air space, a smaller, weaker duckling and hatching problems.

Weighing. Weighing each egg throughout the incubation process is the most accurate way to achieve the proper humidity levels for a successful hatch. Optimally, you want each egg to lose 13% of its weight from hatch to day 25 of the incubation period. More detailed explanations of relative humidity and egg weight loss is beyond the scope of this book, but the specifics can be found both on the Brinsea website at www.brinsea.com and Metzer Farms at www.metzerfarms.com.

Turning. If you are manually turning your eggs, you want to turn them a minimum of five times a day (even more is better, especially the first week) and always an odd number of times – turning 180 degrees side to side each time – so the egg spends every other night on the opposite side. This prevents the developing embryo from sticking to the shell and membrane. Embryos float and rise to the top each time the egg is turned. I set the alarm on my cell phone to remind me to turn the eggs at 6 a.m., 10 a.m., 2 p.m., 6 p.m. and 10 p.m. That schedule works for me; you can set one that works for you, or set the auto turner if your incubator has one. I prefer to turn the eggs manually. I like to think it allows me to bond with the growing embryos and feel more involved in the incubation process, but more likely it's because I'm a bit of a control freak!

Utilize an incubator to maintain a constant temperature and humidity levels for your eggs.

Clean hands! It's very important to wash your hands both prior to and after handling the eggs. Eggshells are extremely porous and bacteria are easily transmitted from your hands through the pores to the developing embryo. If during a candling you see a reddish ring inside the egg, that "blood ring" indicates bacteria have gotten inside the egg and it should be discarded. Contaminated eggs can explode and contaminate other eggs.

Day 5. Five days into the incubation, you should be able to see some "spider" veining when you candle the eggs. The air sac at the blunt end of each egg should have started to expand as well. Being careful not to drop the eggs, work quickly and don't leave the light against the shell for too long because even a few minutes at temperatures above 100°F can kill the embryo. By day 5, the digestive tract, nervous and circulatory systems have formed and the eyes, ears and brain have begun to form. The heart is beating and the legs and tail have begun to develop.

Day 10. By day 10, candling will show significant expansion of the air sac in the blunt end of the egg and a developing embryo. The legs, toes and foot webbing, wings and beak will have begun to form and the bones are beginning to harden. The reproductive system is forming and feathers are beginning to grow. The egg tooth, a nub at the tip of the upper bill which helps the duckling crack the shell and hatch, is beginning to grow. Any eggs not showing any development by day 10 can usually be safely removed as they are most likely infertile or otherwise not going to hatch.

Starting on day 10, the eggs will benefit from daily misting and cooling. Once a day, remove the lid of the incubator and leave it off for 30-60 minutes. The eggs should be let to cool to about 86°F, so they feel neither warm nor cold to the touch. Then mist each egg with a spray bottle of lukewarm water and replace the incubator lid. This mimics a mother duck leaving the nest each day to find something to eat and maybe take a short swim, returning wet to her nest. The misting helps keep the humidity levels high and the membrane moist, which assists the duckling in hatching. The misting also cools the egg surface temperature slightly as the water evaporates. Studies have shown this can greatly improve hatch rates.

Day 26. Continue turning, cooling and misting the eggs as described until day 26. Day 26 is lockdown. At that point, one last candling should be done and any eggs not showing

By day 28, you should see small holes (pips), in the eggs, indicating that the ducklings are starting to work their way out of the shell.

development should be discarded so only viable embryos remain. The eggs should get one last turning, cooling and misting and the incubator closed. The temperature should be decreased to 98.5°F and the humidity should be increased. The eggs should not be touched nor should the incubator be opened until the ducklings have hatched. Opening the incubator causes the humidity level to drop drastically, and moving the eggs can cause them not the hatch. At this point, the ducklings will move into "hatch position" and turning an egg will disorient the duckling, possibly preventing it from being able to successfully break the shell and hatch.

Day 28. Hopefully, if all goes well, on day 28 you will begin to see "pips" (small holes or cracks) appear in the eggshells. The eggs might start rocking and even peep back at you if you quack at them (don't ask me how I know this – just trust me, they will!) The duckling will then begin to make its way out of the shell, "zipping" off the top of the egg and then emerging from the shell.

Once the duckling has made a sizeable hole and is breathing the air in the incubator instead of the air inside the eggshell, it will often take a break to rest up for the final push. This break can last for hours; up to 12 hours is quite common. You shouldn't be tempted to help a duckling unless it has been more than 12 hours since the initial pip and you haven't seen any further

progress, or the duckling is nearly out but seems twisted or wrapped up in the membrane or is "shrink wrapped" in a dried membrane, in which case lightly misting the membrane with warm water can help, as can carefully breaking away pieces of the shell. If you see any bleeding, stop immediately and leave it alone for a few more hours.

The entire hatching process can take 48 hours or longer, so resist the urge to assist and just enjoy watching nature take its course. Leave the ducklings in the incubator until they are rested, dried, fluffy and starting to move around. They don't need to eat or drink for the first 48 hours; they can survive on the nutrients in the egg yolk they absorb just prior to hatching. Once they have dried off, they should be moved to a heated brooder. A few sips of sugar water (3 tablespoons sugar per quart of water) before you put them in the brooder is always a good idea to give your new ducklings some added energy and a good start in life.

After they hatch, ducklings need to stay in the incubator until they are rested, dried and fluffy.

🦆 Hatching Eggs Under a Duck 🦆

Hatching under a duck is far less complicated if you are lucky enough to have a duck cooperate and decide to sit on eggs. The mother duck handles it all. Be sure she has a nice nest of straw in a location safe from the elements, predators and the rest of the flock, and provide her with feed and water close by. Blocking off her nest will ensure that other ducks don't lay more eggs alongside those she is trying to hatch, and will also prevent broken eggs if another duck wants to use the nest to lay her eggs.

Herbs for the nest. It's beneficial to add some fresh or dried herbs to the nest. Oftentimes, ducks aren't overly broody, so some calming herbs such as chamomile, lavender, bee balm, yarrow or lemon balm will relax her and encourage her to sit. Since she's not swimming as much as she would otherwise, you can add some insect-repelling herbs to the nest to help keep it parasite-free; these include basil, bay leaves, mint, rosemary and thyme. Lastly, a few herbs with antibacterial properties are always a good idea. Basil, yarrow and thyme do double-duty here, as does bee balm. Rose petals have antibacterial qualities, plus they smell nice and ducks love to eat them – so toss a few (untreated) rose petals into the nest as well.

The mother duck will turn the eggs, rotate those from the outer ring to the inner ring for even heating, and also most likely kick any non-developing eggs out of the nest. She will control the temperature and humidity. You don't need to candle eggs incubating under a duck. Once the ducklings hatch, she'll keep them warm and teach them how to eat and drink, as well as introduce them to the rest of the flock when they are ready.

Once the ducklings hatch, I move them with the mother duck into a separate cage or crate inside the duck house for a week or so and then let mother duck see how she feels about taking her babies out to explore. Just keep an eye on them initially to be sure the other flock members don't bother them, and be aware that they are very vulnerable to all kinds of predators, including hawks, snakes, rats, cats, and all the other normal predators, so restricting them to an enclosed pen is safest.

When a broody chicken sits on duck eggs

Duck eggs need 28 days to hatch, compared to the 21 days that chicken eggs require, but a broody chicken can, and will, successfully sit on duck eggs for the entire four-week incubation period. To help maintain the proper humidity, duck eggs benefit from putting a piece of sod on the bottom of the nest and misting the eggs once a day. Remember that the ducklings won't have the protective oils a mother duck would impart to their feathers at hatch, so swimming should be limited for the first few weeks. But the look on the mother hen's face when her "baby chicks" hop into the water tub and start bathing will be priceless!

A chicken will sometimes sit on a nest of duck eggs.

THREE

BROODING AND RAISING DUCKLINGS

Raising ducklings, referred to as "brooding," is fairly easy, but be forewarned – it can be messy. Fortunately, ducklings' cuteness more than makes up for any mess they make! They do need a bit of care to survive and grow into healthy ducks, so before bringing your ducklings home, there are some basics you should know.

If you were lucky enough to hatch your ducklings under a duck, she will keep them warm and be sure they learn how to eat and drink; so all you need to do is provide her a secure location out of the elements, separated from the rest of the flock for safety's sake, and safe from predators. It's always a good idea to have a brooding area ready, though, because sometimes a duck will successfully hatch her ducklings but then not be a very good mother, stepping on the ducklings, becoming aggressive toward them, abandoning them or losing interest. In that case, they will need to be moved into a heated brooder.

The brooder box

Whether you hatch fertile eggs in an incubator or buy day-old ducklings locally or online, their first stop will be a heated brooder for the first few weeks of their lives. What kind of brooder should you provide? A simple plastic tote will suffice for the first week or so, as long as it is draft-free and protected from curious children and pets. Ducks are messy and like to play in their water, so a cardboard box generally isn't the best choice for a brooder. They also grow fast. You might prefer to skip the small tote and start them in a large dog crate or cage with cardboard or plastic wrapped around the lower half to prevent drafts (and escapees).

Note: Some reference books recommend brooding ducklings on wire mesh caging to prevent a buildup of spilled water and poop, I don't recommend it. While I have brooded ducklings in a wire rabbit hutch in the past, the wire is very hard on their feet. Duck feet are far more sensitive than chicken feet or rabbit feet. Also, that open wire bottom makes the brooder extremely drafty, which can lead to chilled ducks. If you do use a wire cage, I would suggest leaving only one-third of the floor uncovered, and covering the rest with a solid material such as a piece of wood or rubber yoga mat, etc.

I've found that a bathtub – if you have one to spare – makes a perfect brooder. It's easy to clean and large enough to brood a handful of ducklings until they are ready to be outside. And it's easy to block off the bathroom for the safety of your ducklings.

Or, you could section off space in your garage with a circle of cardboard or chicken wire; cover the floor with a plastic tarp and then a thick layer of chopped straw. It's simple to enlarge as more space is needed. A puppy playpen also makes a wonderful duckling brooder in the garage or a spare room; you can also remove the bottom and move it outside on nice warm days once the ducklings are a few weeks old to give them fresh air and some sunlight.

Whatever you choose to use, you'll need to provide a brooding area for your ducklings that is heated and well ventilated but draft-free. The brooder should be covered or behind closed doors. (If you are brooding outdoors, ensure the brooder is safe from rats, snakes, weasels and other predators.)

A puppy playpen makes a terrific duckling brooder.

18

Space to grow. Since ducklings grow fast, be prepared to provide adequate accommodations. Plan on a minimum of one square foot for every three ducklings at hatch, then for every two ducklings until they are two weeks old, then increase to at least one square foot apiece until they are three weeks old. Of course, more space is always better. After that, they should be able to spend warm, sunny days outside in a safe, enclosed pen, depending on the time of year and your climate, returning to the brooder only to sleep.

A non-slip floor. Ducklings do make a mess. They play in their water and splash it all over the brooder. Since a wet floor can be slippery, it helps to put down rubber shelf liner. The liner provides an easily grippable surface for little duckling feet. If they are brooded on a slippery surface, such as plastic, newspaper or cardboard, your ducklings can end up with foot and leg injuries as they race around, or develop spraddle leg (see page 102).

Rubber shelf liner makes an ideal non-slip floor for ducklings.

Adding litter. After the first few days, once the ducklings have figured out where their food dish is, you can add some litter over the shelf liner.

Good choices include:
- chopped straw
- large pine chips or shavings
- old cotton tee shirts
- even clumps of dirt and grass

Bad choices include:
- newspaper (too slippery, which can lead to spraddle leg)
- sawdust (too dusty, which can lead to respiratory issues) cedar shavings (which can be toxic)
- sand (which, if eaten, can lead to ingested sand-covered poop and impacted crops. Sand, heated by a heat lamp, also can be extremely uncomfortable for small, tender feet.)

Ducklings love to nap amongst dirt and grass clumps.

I like to fill my brooder with dirt clumps. The ducklings love rummaging through the dirt looking for bugs and worms – and the dirt doubles as the grit they need to help them digest the grass that they will inevitably end up eating. If they are eating anything other than chick feed, they do need grit in some form, either small stones/coarse dirt or commercial chick grit. Clumps of grass, dirt and all, give the ducklings something to play with, hide behind and nap on. The clumps of dirt also help build strong immune systems by slowly, and in small amounts, introducing them to the various pathogens and bacteria in the environment in which they will eventually live.

Changing litter. Wet litter should be changed regularly. The moist air in the brooder, warmed by the heat lamp, can quickly lead to mildew and bacteria which can cause aspergillosis in your ducklings, a potentially fatal fungal infection caused by breathing mold spores. If you are using a bathtub brooder, you can greatly reduce the potential for wet litter if you place the water at the drain end – since the water will automatically drain away. After removing the dirty litter, I scrub my brooder down with a white vinegar/water solution (roughly 1 part vinegar to 3 parts water) and let it dry before adding new litter and putting the ducklings back in.

Once they are able to spend days outside, I like to clean the brooder in the morning (after I put them outside) and leave it in the sun to dry all day (if I'm using a tote or crate versus the bathtub, of course!) while the ducks are enjoying play time in their outdoor pen.

Heat

A heat lamp is necessary to keep your ducklings warm for the first few weeks. Be sure to position the heat lamp at one end of the brooder so the ducklings can move away if they get too warm. Watch them for cues. If they are huddled under the lamp and peeping loudly, they are cold; if they are panting or clustered at the far side of their brooder area away from the light, then they are too warm.

An EcoGlow brooder is a good, safe choice for brooder heat for the first week or so, but since ducklings grow so fast, they will quickly be too large to fit underneath.

The brooder temperature should be started at 90 degrees F the first day, then lowered by 7 degrees per week (one degree per day) until they are fully feathered and ready to go outside full time – at about 6 weeks old, weather dependent.

An EcoGlo brooder provides heat for young ducklings.

Brooder Temperature Chart

Reduce the temperature one degree per day (shown in Fahrenheit):

1st week 90-84°	4th week 69-63°
2nd week 83-77°	5th week 62-56°
3rd week 76-70°	6th week 55°

At 2-3 weeks old, your ducklings should be able to spend some time outside on warm, sunny days. It's very important that ducklings get fresh air, sunlight and plenty of exercise, as well as exposure to fresh grass, for healthy development and growth. Of course, this is dependent on your climate – and they need to be protected from predators at all times, including wild birds, crows and snakes.

For spring hatches: at 6-8 weeks old, once your ducklings are fully feathered, they should be ready to move outside and away from the heat permanently – into a dry, draft-free house.

Water

When you first bring your ducklings home (or take them out of the incubator), dip each duckling's bill into a dish of sugar water (1/3 cup sugar per gallon of water) for an added energy boost. You can give them sugar water for the first few days, then switch to plain water. Room temperature or lukewarm water is best. Position the waterer at the opposite end of the brooder from the heat lamp so that splashing water doesn't shatter the bulb.

Adding a splash of apple cider vinegar to the water every few days is also extremely beneficial as a health and immune system booster and thought to help prevent internal worms, bacterial infection and coccidiosis. Be sure not to use metal if you are adding apple cider vinegar, because it will rust even galvanized metal waterers.

What kind of waterer? A traditional Mason jar chick waterer doesn't work well for ducklings. The design is too unstable and tips over too easily around rambunctious, clumsy ducklings. Ducklings are also able to empty a fountain (or gravity) waterer in mere minutes. Instead, I like to use a shallow, flat-bottomed dish or stoneware pie plate. Setting the water dish on a rimmed cookie sheet can help keep the water mess contained.

Ducklings need water at all times, day and night. They are prone to choking if they don't have access to water any time they are eating, so never leave feed for them unless they have ample water as well to help them swallow their feed. They also need to be able to keep their nostril membranes moist, clear of feed and debris and to clean their eyes in the water. As they grow, they will need progressively deeper water containers, but they can easily drown or get chilled if they sit in the water, so add some stones to help prevent accidental drowning – and be sure the water is just deep enough so that they can submerge their bill and head.

By 2 weeks old, your ducklings will start to preen and activate their oil glands, which help to waterproof their feathers. By providing them a water dish

Setting a waterer in a cookie sheet helps eliminate messes.

2 or 3 inches deep, you can facilitate that process for them. By choosing a slightly deeper water container with a smaller diameter you can prevent them sitting in it for the most part.

Ducklings are drinkers! A week-old duckling will drink about half a gallon of water a week. By the time a duckling is seven weeks old, it will drink almost a half a gallon of water a DAY, so be sure there is always clean, fresh water available. Water should be changed at least daily and very possibly several times a day if you are able, to ensure debris-free, clean water. Remember though, ducklings are messy, so

Provide ducklings with fresh water every day.

if you expect crystal clear water, your expectations are too high! Feed, some dirt or straw or shavings in the water are not a problem; but feces are.

Feed

Feed should be provided at all times for your ducklings, away from the heat source and fairly close to the water. Ducklings can choke if they eat without drinking, so you want to keep the feeder and waterer next to each other. Traditional Mason jar chick feeders don't work well for ducklings any better than the Mason jar chick waterers mentioned above, since the little guys have trouble fitting their bills in the small holes – and the Mason jar feeders are also prone to tipping over. Instead, I use a small terracotta planter base or a sturdy casserole dish for feed. Since ducklings go back and forth between their feed and water, the terracotta also helps keep the feed a bit more dry by wicking moisture away. Leftover feed should always be discarded at the end of each day to prevent mold, but wet feed can be left during the day without a problem.

What kind of feed? Regular chick feed is fine for ducklings if you can't find waterfowl starter feed. However, ducklings should only be fed UN-MEDICATED feed. They eat more feed per ounce of body weight than chicks, so it's possible for them to over-medicate themselves, and since they aren't as susceptible to coccidiosis anyway, there is no need to feed them medicated feed.

- For the first 2-3 weeks: Offer chick feed (19-21% protein) all day long, and preferably through the night as well (as mentioned previously, just be sure ample water is also provided).
- From 3-9 weeks: You will notice a substantial growth spurt at about 3 weeks, so switch at that point to the lower protein starter/grower feed (15-17% protein) until they are about 9 weeks old. Too much protein can cause leg and wing deformities, kidney and liver damage.
- From 9-20 weeks: Ducks are close to their adult weight by 9 weeks and grow slowly between 9-20 weeks, so give them grower feed (15% protein) – or continue to feed them the starter/grower feed during this phase.
- Around 20 weeks or so: At this point they will be ready for layer feed (16% protein with added calcium) to prepare them to begin laying. Ducks should start laying any time after 21 weeks old. A good layer will lay for more than four years and live to be eight to ten years old on average, although ducks living a dozen to 15 years is not unheard of.

Feed supplements

Brewer's yeast. Ducklings require more niacin than chick feed provides, to help with bone growth, especially in their legs, so you need to add brewer's yeast to their diet. Too little niacin can cause leg deformities and weakness and also result in smaller ducks than normal. It's important to supplement them with niacin, especially during the starter/grower phase.

Garlic. Garlic is another beneficial addition to ducklings' diets. Fresh or powdered, garlic improves general health, boosts immune systems and supports respiratory health. When the ducks are older, I crush whole fresh garlic cloves into their water every few weeks, but for ducklings, the powdered form is less of a choking hazard.

To combine both of these beneficial supplements, I use a garlic powder/brewer's yeast mix from Thomas Laboratories made specifically for poultry. A 2.5% ratio of the mix in the feed is the recommended dosage. I find that a sprinkle over the top of the feed dish each time I refill it is easier than trying to measure out an exact amount. If you are not seeing any issues with standing or walking, sprinkling the powder, while less precise, is far easier and most likely sufficient. On the flip side, if your ducklings aren't growing well or

Your ducklings' feed should be unmedicated.

have trouble with their legs, adding a bit more can help. You can also premix in bulk, mixing 1/2 cup of the powder into a 20-pound bag of feed. Just be sure to remix the feed before you measure it out since the powders will tend to settle at the bottom.

Probiotic powder. This is another excellent feed supplement. It aids in intestinal tract health and helps maintain a healthy digestive system. Just a sprinkle in their feed will provide numerous benefits. A ratio of 2% probiotic powder to feed is generally recommended. Again, guesstimate the amount or premix in bulk.

Oats. In addition to the brewer's yeast and probiotic powder, you can add raw, uncooked oats to the feed, gradually increasing to a 25% oat to 75% feed ratio. This can help avoid a condition called angel wing, which is caused by too much protein (see page 90 for more information).

Treats

Treats help keep ducklings busy and also provide a varied diet. I have found that ducklings who are fed a wide variety of vegetable scraps, herbs and weeds are more apt to be willing to try new foods as adults. As with adult ducks, duckling treats should be limited and include healthy choices, such as kale, Swiss chard, tender chopped grass, herbs, weeds, and dandelion greens. These are all excellent sources of nutrients.

A handful of chopped greens or peas in a dish of water is a great way to amuse your ducklings and keep them occupied. In general, anything green will be a huge hit with them. *Note:* Iceberg lettuce should be avoided because it has little nutritional value and can cause diarrhea. (For a listing of edible plants see the Appendix, page 136.)

Ducklings can be fairly picky about treats, but worms, watermelon, corn, scrambled eggs, oatmeal, and halved grapes all are favorites. Fresh minced garlic is a wonderful addition to their diet because it helps build strong immune systems. Culinary herbs and edible flowers are also incredibly nutritious. A very healthy treat that my ducklings love is scrambled eggs mixed with ground wheat, raw oats, some powdered milk and brewer's yeast.

Slowly introduce your ducklings to new foods, making sure that starter feed still makes up the majority of their diet. If ducklings develop a taste for healthy treats early on, they are more likely to continue to eat them as they get older. In addition to the grass clumps I put in their brooder, I also offer just-a-few-days-old ducklings chopped herbs, weeds and edible flowers right from the start – figuring that ducklings hatched outside under a mother duck

It's important to bond with ducks while they're young.

would have access to all these treats as soon as they left the nest. Please remember that any time ducklings are given anything to eat other than commercial chick feed, they need grit to help them digest the food. Coarse dirt or commercial chick grit are both fine.

Quality time with your ducklings. Hand feed your ducklings their treats and talk with them. Handle your ducklings often (be sure small children are always supervised and taught how to hold the ducklings to avoid injury). Ducklings will imprint on you and become far friendlier adult ducks the more time you spend with them as ducklings.

Swimming

Ducklings can swim at birth, but those hatched in an incubator instead of under a mother hen won't have the benefit of her oils to waterproof them. Although your ducklings will start to preen and oil their feathers around 2 weeks old, they won't have fully developed the oils on their feathers until they are about a month old. Swims should be short and supervised until that time, because your ducklings can get waterlogged and chilled and easily drown. Drowning is the number one cause of fatalities in backyard ducklings.

Ducklings' swim time should be supervised.

However, letting your ducklings swim in warm water in a rubber washtub, sink or bathtub for a few minutes at a time and then drying them off carefully with a towel before returning them to the heated brooder will help teach them to preen and develop their preen gland – which is how they waterproof their feathers. Never leave young ducklings in any water unsupervised, not that you would be able to anyway! Watching ducklings paddle around and dive under the surface of the water is one of the most adorable things I've ever seen and I would be hard pressed to walk away!

By the time your ducklings are about a month old, they will be able to swim on their own, but be sure they can easily get in and out of their pool or tub by providing a ramp or stepping stones.

The big outdoors

At 2-3 weeks old, ducklings can be outside for short periods of time on warm, sunny days with adequate protection from predators, sun and rain. Proper precautions should be taken whenever your ducklings are outdoors as they are not able to move quickly, so they are easy prey for predators. I use a rabbit hutch for them to get fresh air and then allow them to roam the grass inside a large pen made of 1/2" welded wire. As they get older and the weather gets warmer, their time outdoors can be lengthened from minutes to several hours. Be sure to provide shade and plenty of water for them.

By 6 weeks old, your ducklings should be fully feathered and able to be outdoors permanently, as long as temperatures don't drop below 50° F. Their house should be predator-proof, dry and out of the wind. (See Chapter 7.) Straw makes a nice choice for bedding since it provides excellent insulation and warmth. Ducks don't perch on roosts like chickens; they will just settle on the floor in the straw, so be sure there is a nice thick layer of bedding for them to burrow in at night.

If you are introducing your ducklings into a run with other ducks or chickens, you should keep them in a small pen inside the run for a week or two, so the two groups can get used to each other slowly. You want to minimize squabbles, allow the ducklings to get a bit bigger, and avoid any other integration issues.

🦆 Starter Supplies: a Quick Checklist 🦆

Here's what you'll need before your ducklings arrive:

- *Brooder box*
- *Heat lamp with red bulb (plus spare bulb) – or EcoGlow brooder*
- *Starter feed*
- *Brewer's yeast*
- *Raw oats*
- *Small, shallow dishes for feed and water*
- *Sugar*
- *Chick-sized grit or clumps of dirt*
- *Rubber shelf liner*
- *Straw or other bedding*

Determining the sex of your ducklings

Determining the sex of a duckling can be done when they are a few days old.

At hatch: Ducklings' boy and girl parts are inside their bodies, so determining their sex isn't quite as easy as it is with humans or various other types of animals, but it is easier than trying to sex baby chicks. However, a quick look between the legs isn't going to tell you anything! Instead, an examination of the vent and sex organs inside it must be done. Ducklings are able to be sexed at hatch and anytime thereafter, but can easily be injured, so best not to attempt vent sexing unless you are experienced. It is recommended to sex them when they are only a few days old; at that age a duckling is small enough to hold in one hand on its back while you gently pull down on the sides of the vent at the base of the tail to see if a tiny penis emerges or not. Note that false positives for females are common until you get good at sexing. If you're not confident you can get it right, then you should have someone with experience do it for you or show you how. Or you can always just wait to see who eventually starts laying eggs.

At 2-3 weeks: Males will generally be larger and taller.

Around 5 weeks: Females will start quacking, while males will still peep for another few weeks.

At 5-6 weeks: Colored breeds may begin to show their colors (i.e., green head on a Mallard drake).

By 6-8 weeks: Females will make a loud quack, males will have a deeper, raspy quack.

By 16 weeks: Males will be sporting a curly tail feather.

By 21 weeks (but could be as late as 25 weeks or longer): Females will start laying and their feet and bills will fade to a pale orange, while the drakes will retain their dark orange bills and feet. Why do the female layers have paler bills? It's because of xyanthophyll, the compound that not only colors duck bills and feet, but also egg yolks – so laying ducks funnel the majority of their pigment to the egg yolk, leaving very little for their own coloring.

FOUR

A DAY IN A LIFE WITH DUCKS

Ducks love routines and hate to be stressed. Any change in routine can reduce their productivity and the number of eggs they lay. Remember that "routine" is what they are used to, so a flock that lives in a pen with barking dogs as next door neighbors will get used to that noise, or a flock living where airplanes regularly fly overhead will consider that part of their routine. However, sudden or unexpected noises or actions will stress them.

I keep to a pretty strict routine with our ducks because I can, since I work from home. But you can adapt your routine to your own work or family schedule – even if no one is home during the day (see page 46). Just pick a routine and stick to it and your ducks will be happy. And get the whole family involved!

Collecting eggs is a part of the daily morning routine.

5:30 A.M. · 8:00 A.M.
BREAKFAST FEEDING/TURNOUT
(takes about 10-20 minutes)

Happy ducks enjoy a variety of breakfast foods, including oats, corn and dried berries.

As the sun peeks up over the tops of the trees, I finish the last of my coffee and get dressed in my "barn attire." My schedule (and the ducks') isn't dictated by the time on the clock on the kitchen wall, but entirely on the sun. They want to be let out at sunup, whether that comes at 5:30 a.m. or 8:00 a.m., so in the summer I might be out there letting them out as soon as my husband leaves for work, while in the dead of winter I can relax and enjoy a cup of coffee or two before heading down to the duck house.

In the warm months, I usually just throw on a tee shirt and pair of capris, along with some rubber clogs (washable, rinseable footwear is a must with ducks) – but this morning because it's winter, I bundle up in long johns, jeans, warm wool socks, a fleece, and then put on my barn jacket and gloves. I pull on a pair of muck boots and head out.

It's cold outside, so I've heated up some water in the teakettle and have a special treat for the ducks to go with their breakfast. They're getting a pan of oats, cracked corn, dried cranberries and mealworms, moistened with the warm water. Your ducks should all waddle happily out of their house, ready to start a new day. This is a great time to give each a quick visual check for any limping, lethargy or other odd behavior.

Feed

Since I raise my chickens and ducks together, my ducks' main feed is an organic poultry layer feed, to which I add a few natural supplements (see below). I pre-mix one or two bags of feed in a large rubber tub with all the supplements and then dole out about a week's worth into a covered metal container, with a plastic pail set inside from which to feed. I try to not mix more than a 3-week supply at a time and keep the feed in a dry, dark, cool place to reduce the loss of nutrients and prevent any mold from forming. Ducks are extremely susceptible to mold toxins, so any bad-smelling feed that you suspect is spoiled or contaminated should be thrown out immediately.

Store your duck feed in a cool, dark place.

Can I use chicken feed for ducks?

Chicken feed isn't a perfect match for the slightly different protein, niacin, and calcium needs of ducks; but if you can't locate a source for waterfowl feed or raise both chickens and ducks, you most certainly can feed your ducks chicken layer feed once they are at least 17-18 weeks old and approaching laying age.

A grown duck will eat approximately 4-6 ounces of feed a day, so I measure out about 1/2 cup of feed per duck in the morning and they eat it throughout the day. I will refill their feeders in late afternoon if needed, and also give them lots of leafy greens and other healthy treats every day in addition to allowing them a bit of free-range time just before dusk so they can find worms, bugs and other yummy things to eat in the pasture.

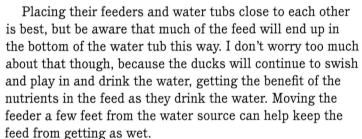

When they eat, ducks grab a bill full of feed and then dunk their head into their water, swishing it all around; this moistens the food to allow them to swallow it without choking. This means two things:

1) Ducks should never have feed without fresh water nearby.

2) Gravity feeders commonly used for chickens don't work because the ducks get the feed all wet and the feeder will get clogged. Ducks also have a knack for being able to empty a gravity feeder in two minutes flat just for fun and you'll have feed all over the ground. Instead, I scour thrift shops and yard sales and buy large stoneware casserole dishes or other sturdy containers to use for the ducks' feed. They are easy to clean and refill, extremely affordable and sturdy enough that the ducks can't tip them over.

Have plenty of water close by to their feed during mealtime.

Placing their feeders and water tubs close to each other is best, but be aware that much of the feed will end up in the bottom of the water tub this way. I don't worry too much about that though, because the ducks will continue to swish and play in and drink the water, getting the benefit of the nutrients in the feed as they drink the water. Moving the feeder a few feet from the water source can help keep the feed from getting as wet.

Water

I give our ducks a water tub, versus a gravity waterer or nipple waterer. As with the gravity feeders, ducks see gravity and nipple waterers as a challenge to see how fast they can empty them. Once the ducks learn they can make water come out of the nipple every time they poke at it, it becomes such a fun game for them!

The rubber tubs are sturdy and easy to clean and refill. There are various sizes available. The larger tubs will tempt the ducks to try and swim in them, so I have several sizes – but all are at least 6" deep, and some of the larger ones are 8" deep to ensure the ducks can dunk their bills and heads completely.

"like a duck to water…" Unlike chickens (and most other livestock), who view water as merely something to drink, ducks consider a tub of water more all-encompassing; certainly as a source of hydration, but also a place to bathe, play, search for food and sit in to keep cool in the summer. It's also how they keep their nostrils free of caked feed, mud and debris and their nostril membranes moist. As a result, you will find yourself changing your ducks' water more often than chicken water, if you have chickens. There's really no way of getting around the ducks' making their water tubs a mucky mess – or if there is, I haven't discovered it yet!

Make sure that any vessel that you use for your ducks' water be deep enough to dip their bills into it while eating.

Ducks need to have access to water deep enough to dip their entire bill into while they are eating to ensure they can drink enough to get the feed down and also keep their bill vents (nostrils) clear of feed debris. It's pretty much inevitable that your ducks will also sit in and muddy up their water as soon as you refill it. You will drive yourself nuts trying to keep your ducks' water crystal clear. Remember, the water needs to be fresh. And remember that with ducks, fresh and clear are not synonymous. Mud and debris in the water (and you wouldn't believe some of the stuff I've found in my ducks' water tubs, including pine cones, rocks, eggs – yup eggs – and even a tennis ball once) is fine; poop and fecal matter is not.

So changing the ducks' water at least twice a day is probably the minimum you can get away with, and if you can dump it out and refill it more often, that's great but not completely necessary. You don't want stagnant water to sit for days, however. That will attract mosquitoes and can allow bacteria to grow, which can cause botulism and can be fatal to ducks. But bottom line, don't stress over muddy water. Even our chickens seem to have gotten used to drinking it and I like to think that the dirt and mud adds some essential nutrients to the water.

Apple Cider Vinegar for Cleaner Water

To help keep the water cleaner and free of algae and bacteria, add a tablespoon of apple cider vinegar to the water a few times a week. It also has wonderful health benefits for the ducks, including improving their digestive system, boosting their immune system and increasing the absorption of calcium and other nutrients in their food. Apple cider vinegar improves mucus membrane health, and as an antiseptic, helps kills germs that cause respiratory illness.

Providing grit. Ducks don't exactly have a "crop" like chickens do, but their esophagus expands to accommodate the food they eat before it's digested. They store grit (small stones, pebbles or coarse dirt) in their gizzard, which helps break down the food. If your ducks don't have regular access to new soil, you will need to provide them a dish of commercial grit – available at your feed store – to nibble at free-choice so they can eat as much or as little as they need.

The quality of a duck's eggshell is determined by their feed and supplements.

Egg shells for calcium. Even if your ducks are eating a good quality layer feed, I recommend you also provide your ducks crushed eggshell or commercially purchased oyster shell free-choice in a dish as well; this way your ducks can supplement the calcium in the feed and other foods they eat to ensure nice hard eggshells. Save your eggshells by rinsing them, removing the membrane, then air-drying them. Once dry, store them in an open container on the kitchen counter until you have a good amount, then crush them to about 1/8" pieces with your fingers or a rolling pin.

Oats: Oatmeal is an excellent source of vitamins and minerals and ducks love it. Oats help lower the protein content of the feed – too much protein can cause too-rapid growth in ducks and a condition called angel wing. I add one large grocery store-size canister of old-fashioned oats to a 40-50 lb. bag of layer feed.

Probiotic powder: Adding probiotics to your ducks' diet is like adding yogurt to your own diet. Probiotics aid digestion, assure better intestinal health and support the immune system in general. The addition of probiotics can assist in nutrient absorption and result in larger eggs with stronger shells. Probiotics are also thought to help combat Salmon-

Enhance the nutritional value of your ducks' diet by adding feed supplements.

ella and other pathogens in flocks, reducing the incidence of Salmonella, specifically, by 99% in flocks whose diet includes probiotics. I add some probiotic powder to a bag of layer feed (check probiotic powder bag for recommended amount).

Seaweed/Kelp: Kelp is also rich in nutrients and a natural fertility booster, as well as overall health booster. Some studies show that adding seaweed to a bird's diet can reduce the cholesterol levels and increase the omega-3s in their eggs. The prebiotics in kelp work with probiotics to increase the levels of good bacteria in the digestive tract. I add some dried sea kelp to the 40-50 lb. bag of feed (check sea kelp bag for recommended amount).

Brewer's yeast: Brewer's yeast is probably the most beneficial supplement you can give your ducks. In addition to containing all of the essential amino acids and supporting the circulatory, nervous and cardiovascular system, it provides ducks the niacin they need to grow nice strong legs and bones. Regular layer feed formulated for chickens generally doesn't contain a high enough level of brewer's yeast for ducks.

Garlic: Adding garlic to your ducks' diet will result in better feed conversion, better overall health, stronger respiratory and immune systems and reduced manure odor. Garlic is also thought to be a natural wormer and flea and tick repellent. You can add fresh garlic cloves to your ducks' water but they tend to just fish them out and eat them, so I prefer to add garlic powder to the feed to guarantee that everyone gets the same amount (and not just a greedy few!).

Note: The Fresh Eggs Daily line of poultry feed supplements has been formulated with backyard ducks in mind and includes Brewers Yeast with Garlic, Coop Kelp and Poultry Probiotics and is available from Amazon.com.

Dried herbs: I also like to grow culinary herbs to feed to our ducks. In the summer, I pick the herbs straight from the garden to feed them, and I dry some excess to crush and add to their feed in the winter. I mix whatever I have extra of but I always like to include these favorites:

- *Laying stimulants:* Fennel, marjoram, nasturtium, parsley
- *Respiratory health:* Bee balm, dill, oregano, thyme
- *Overall health:* Cilantro, oregano, sage, tarragon
- *For orange egg yolks, feet and bills:* Alfalfa, basil, dandelion, marigolds

Treat your ducks to nutrient-rich greens like these dandelions.

COLLECTING EGGS AND SPRUCING UP THE DUCK HOUSE
(takes about 5 minutes)

After I let the ducks out and fill the feeders and waterers so the ducks are chattering contentedly instead of quacking at me to LET THEM OUT!!!! – it's time to check for eggs and spruce up the duck house.

Ducks usually lay their eggs an hour or two before dawn, or just after, and like to cover them up with the nesting material to hide them. I generally wait until the sun is coming up to let the ducks out because if you let your ducks out before sunup, they might lay their eggs outside instead of in the duck house; so you will often find eggs lying around the pen in the mud, or even at the bottom of the duck pool – and each morning can become a bit of an Easter egg hunt.

Kids love to collect eggs!

Nesting boxes. These aren't necessary since your ducks will most likely make their own nests in the straw in the corners of their house (in fact, that's where I found a half-dozen eggs this morning), but if you do build floor-level nesting boxes, they should be 15" x 15" and filled with soft bedding such as pine chips or chopped straw. A 3-4" board across the front helps prevent the ducks from kicking all their nesting material out of the boxes. You can also use wooden boxes or plastic bins or totes.

Most breeds of ducks will lay consistently throughout the year, even during the shorter days of winter without any additional light. Like chickens, they lay one egg every day or so, and will slow or stop while they are molting or broody, although most domestic duck breeds haven't retained the tendency to go broody. Many duck breeds outlay chickens, and with proper diet and care can lay well for 5-6 years.

In the duck house. After I have collected the eggs, I remove any dirty or poopy straw from the duck house and usually toss it over any muddy spots in the run to absorb some of the muck. Duck poop is nearly 90% water, so it's pretty wet. I find straw in the duck house to be the best litter choice. It is drier than hay, less likely to mold and less dusty than shavings. It also is a very good insulator. In the winter, a nice thick layer of straw 12" or more helps keep your ducks warm at night.

I don't leave feed or water inside the duck house at night; this helps keep the house a lot drier and cleaner. However, since ducks are fairly active at night, I do have a small night pen attached to our duck house inside the larger run and I put any leftover feed from each day in there, along with a tub of water and let the ducks finish it off. This works well because there is never any wasted feed at the end of they day, it keeps the ducks busy overnight, and since the ducks eat throughout the night, they are less apt to fight the chickens (who share the run with them) for breakfast in the morning.

Leave the duck house door open into the pen.

Part of the night pen is covered, but I find that ducks prefer to sleep outside except on the coldest, most blustery of nights. I leave the duck house door open into the pen, so the ducks can come and go as they please, unless sleet or freezing rain is predicted – then I do shut them up in their house.

10:00 A.M.
FILLING THE POOL
(takes about 10-15 minutes)

It's a sunny day today, so even though it's the middle of winter, I have decided to fill a pool for the ducks. In the summer, I fill a kiddie pool for them every day, in the colder months I use an old horse trough and try to give them some pool time at least once a week. If you live where it's extremely cold for long periods of time and a pool would just freeze too quickly, the ducks will be fine with a deep tub to take a quick dip in at least every other week or so. Ducks are extremely cold-hardy and have often stood patiently waiting for me to come and break the ice on the top of their pool. They will

Ducks will swim year-round – even if it's cold out!

swim in temperatures well below freezing, but I generally limit pool time in the winter mainly because the water line to our barn freezes – and much as I love the ducks, carrying buckets of warm water from the house to fill a pool for them really would test my limits!

Bottom line: as often as you can or will fill a pool for the ducks, they will use it. It doesn't have to be every day, but it can be year-round – that's up to you. See Chapter 8 for some duck pool ideas.

I sit and watch the ducks as the pool is filling. They rarely wait until the entire pool is full and usually hop right into the empty pool and enjoy playing in the hose spray. I enjoy watching them so much that I most often stay long after the pool is full, just appreciating their utter joy with swimming.

Cover your duck pen to protect against predators.

The duck pen. Although our ducks don't stray far from their pool or water tubs, your duck pen should allow at least 8-12 square feet per duck as a minimum – of course, the larger the better – and allow plenty of room for the feed, water, pool area, plus room to exercise, and shady areas for nap time.

My pen is large enough that I can section off areas and grow grass seed for the ducks. I rotate them back and forth once the grass has grown to a few inches tall. Winter rye, barley, millet and other grains are big favorites for the ducks and allow our non-free-range ducks a faux free-range diet.

Note: I don't recommend free-ranging your ducks unless you are home all day and have specially trained livestock guard dogs. Ducks are extremely slow and unwieldy on the ground and (other than Mallards and Muscovies) can't fly, so they are exceptionally vulnerable to predator attacks. A large, safe pen is your best bet.

Your duck pen, whether the ducks are locked up safely at night or not, should be made of chain link or welded wire with the fencing sunk at least a foot or so into the ground to deter digging predators. The pen should have chicken wire or smaller gauge wire wrapped around the lower 2 feet or so to deter pesky raccoons and other predators from reaching in through the fencing and grabbing unsuspecting ducks, who for some reason seem to prefer snoozing right up against the pen fencing. The pen should also be covered to prevent foxes or raccoons from climbing the sides and also to protect from aerial predators such as hawks, owls and other raptors flying in.

Deterring predators. Depending on where you live, dogs, foxes, raccoons, weasels (minks or fisher cats) and coyotes will be the greatest threat to your ducks. While ducks are most vulnerable at night, dogs are a huge daytime threat. Foxes will hunt at dusk or dawn for the most part, but sometimes will also come out on rainy or cloudy dark days, especially during breeding season. Coyotes do hunt by day, while raccoons mainly hunt by night. Keeping underbrush around your coop and pen cleared can help keep predators at bay by not providing them places to hide or sneak up. But planting foliage and landscaping right around your pen area can actually help deter predators by providing a visual block of your flock. Planting some shrubs or small bushes around the inside perimeter of your run is a great idea also, not only to shield your ducks from predators' eyes but also provide them a nice place to nap. Since ducks only tend to sleep for several hours a night, they enjoy taking a long nap each afternoon, most often tucked underneath a bush.

2:00 P.M.
AFTERNOON TREATS
(takes about 5 minutes)

I've had my lunch and now it's time to bring the ducks some lunch too. Ducks put weight on fast if they are overfed, or fed unhealthy foods (just like humans!). This can stress their feet and legs, so treats should be fed sparingly and you should stick to mostly healthy treats – fresh greens, weeds, whole grains, fruits and the like. I keep a container of kitchen scraps in the refrigerator for them. Today's treat is broccoli stalks, Brussels sprouts and some cut up tomatoes, along with some ends of garlic cloves and ginger peel. (See Chapter 6 for a more complete list of healthy duck treats.)

I try to keep treats to about 10% of the ducks' total diet, although I don't count grass, weeds or leafy greens as treats since they are so healthy and a normal part of a wild duck's diet. These are considered "green" treats and can be fed pretty much in unlimited amounts.

Now that the ducks have filled up on their layer feed, I let them indulge a bit on the treats. I also will replenish their feed dishes if necessary and usually dump and refill their water tubs.

Treat your ducks to plenty of greens and berries.

43

I make it a point to check the run for debris. It's amazing what the ducks can dig up! Since ducks don't tend to look where they are walking, and their feet are more tender than chicken feet, it's a good idea to periodically check your pen for anything sharp that could cut into the ducks' feet, such as pieces of glass or metal, screws, nails, even sharp sticks, stones or pine cones.

3:00 P.M. – 9:00 P.M.
FREE-RANGE TIME
(1-2 hours)

Let ducks roam later in the day, when the threat of predators is lower.

Free-range time is more dictated by the hours of daylight and the weather than the clock. About an hour or two before dusk, I let the ducks out of the run and out into the pasture. This is my favorite time of day and I know it's theirs also. By now, the hawks are pretty much absent and not a threat, but because of other potential predators, I stay outside with the ducks. Usually our dogs are also outside, and in nice weather my husband and I will often spend late afternoons out on the back patio, watching the ducks roam the pasture, looking for bugs and worms, splashing around in any puddles and having fun stretching their legs and exploring.

While the ducks are out roaming, I put any leftover feed from the day into the ducks' night pen along with a fresh tub of water, so they can eat and drink. This works out rather well, ensuring there is never any wasted feed – but also, the following morning the ducks are less likely to fight over the feed with the chickens, since they (the ducks) have already been eating long into the night.

5:00 P.M. – 9:00 P.M.
BEDTIME

Above, all, keep your duck-keeping fun!

Again, bedtime is dictated by the position of the sun rather than the time on the clock. In the summer, the ducks are often still out playing in their pool until nearly 10 p.m., while in the dead of winter, I can get them locked up by 4:30 p.m. Our own dinner time revolves around the time the ducks go in, and dinner invitations are always determined depending on the time of year. In the summer, we go out early (no free-ranging those evenings) and are sure to be home by dark; in the winter, we lock the ducks up and then head out to dinner. A minor inconvenience, but most of our good friends understand, and holiday meals have even been planned based on the ducks' bedtime!

Ducks don't have a strong pecking order; that's not to say that they don't have a strong flock instinct. You will rarely see a duck all by itself. They travel in a group, eat together, nap together, preen together and swim together. This makes it fairly easy to "herd" them or train them to return to their house at night. Unlike chickens (who I find easier to call or shake a treat canister for and encourage to follow me into the coop), ducks are easier herded from behind. Using a long bamboo stick, I just get one duck heading in the right direction and generally the rest will happily follow.

Ducks will tend to make noise if they are bothered by lights through the night. Moving lights can be a stressor as well, so be sure to position your duck house and pen where lights from passing cars, security lights or motion-detecting lights won't bother them at night or you might have some pretty unhappy neighbors.

Now that the ducks are tucked in safely, I can think about getting our dinner ready and then relaxing with a good book, or watching a movie and doing some knitting.

Next day, repeat.

WHEN NO ONE IS HOME DURING THE DAY
(Simple Duck-Care Strategy)

You might think you can't keep ducks if the parents work, the kids are in school and no one is home all day, but actually ducks can fit quite well into a working family.

Weekdays before leaving for work or school:
- Let the ducks out of their house into a roomy, attached predator-proof pen.
- Fill their feeder (allocate 1-2 cups of feed per duck) and fill several water tubs in case they spill one.
- Collect the eggs. Ducks generally lay their eggs pre-dawn, so by the time you let them out, they should be done laying.

Weekdays when you get home:
- Give the ducks some supervised free-range time in the yard if possible.
- If no free-range (i.e., no one can stay outside with them), at least give them some treats in their pen.
- Refill feeders so your ducks can eat again before bedtime.
- Rinse and refill waterers.
- Clean and refill their pool if you have time.
- Lock the ducks up at dusk.

They'll love it if you add some rose petals, peas or other treats to their water.

Weekends:

- Scrub out feeders and waterers.
- Dump out their pool, scrub it and refill it.
- Supervise some free-range time.
- Refresh bedding in their house, or do a full cleanout if necessary, letting the house air out and then replacing the bedding with new litter before dusk.
- Do your duck check-up, focusing especially on feet for cuts or bumblefoot.

Come on in – the water's great!

As long as the ducks are confined to a safe pen for the day, they will be just fine without anyone being home. Just be sure the pen is completely enclosed, covered too, and made of 1" or smaller welded wire fencing which is sunk into the ground. Be sure the door has a predator-proof lock – a padlock is safest if you're worried about the human kind of predators or neighborhood kids letting the ducks out (sadly, that happens).

In the summer. When temperatures rise, providing a kiddie pool is important to allow the ducks to cool off and get a drink in case they tip over their water tubs. Be sure they have enough water to last all day, set in the shade to keep it cooler. And be sure the door to their duck house can't blow shut, so they can come and go as they please to get out of the sun or wind if they wish.

Treats. Some treats can help relieve boredom, so to help keep them busy while you're gone, give them hanging baskets filled with leafy greens, a head of cabbage for them to roll around the pen, broccoli stalks or halved squash or pumpkins, or even a pile of leaves or straw. That should keep them out of trouble (hopefully!).

FIVE

DUCK BEHAVIOR

Ducks are funny, smart, inquisitive, quirky, fascinating little creatures. Much of their behavior is instinctive and embedded in their DNA despite having been domesticated since ancient times. I don't find their behavior to necessarily be breed specific. Our ducks all seem to communicate with each other in much the same way, regardless of breed, and everyone seems to understand – which makes sense, since every domestic breed known today descended from the Mallard (except for the Muscovy, which descended from the wood duck).

Having spent countless hours watching my ducks, and trying to decipher the different things they do, then reading as much as I could relating to both domestic and wild duck behavior, I believe I have figured out what much of their behavior means – although I'm not sure the chickens always understand the ducks' message! Our ducks have a bad habit of sneaking up and "goosing" the chickens and then scampering away (It's very clear to me where THAT term came from!).

So when your ducks do this, it means…

Head bobbing

 You will notice lots of head bobbing within your flock. The ducks sort of make a soft chattering noise and bob their heads up and down at each other. Often occurring during prime mating season and during duck "courtship" in the spring, the head bobs are a duck's way of flirting. Head bobs are also a sign of submission to a more dominant duck. The duck lower on the totem pole will bob her head down low to the dominant duck – who will often return the bob, accepting the other duck. Head bobs are also used to welcome a new flock member. If I sit and hold or pet one of our ducks for awhile, then put her back down, she'll head back to the rest of the ducks as fast as she can, bobbing her head the whole way, and the rest will also start their heads bobbing, welcoming her back. Our ducks also bob their heads when they get agitated about something or if I start filling their pool or they spot me with some treats. So I guess you could say head bobbing signifies an excited duck, in more ways than one!

Head tilting

Vision is a duck's most important sense. A duck can see two to three times farther than a human can. This is critical to survival, allowing them to sight a predator from afar to give them as much advance notice as possible. With eyes set to the sides of the head and fitted into the socket – unlike those of humans and mammals which can rotate within the socket – ducks have to actually move or tilt their heads in order to see in a different direction. They are always conscious of things moving overhead, so head tilting is done to scan the sky for predators. Our ducks are super vigilant about hawks and owls, and will tilt their heads and stand frozen in place until the threat passes; they have even been known to freeze at the sight of an airplane passing overhead until it disappears out of sight. Butterflies and falling leaves have also been cause for concern at times. Regardless, if you catch your ducks tilting their heads to one side, it's almost guaranteed there's something up in the sky, whether or not you can spot it.

Walking in a row

Having all your ducks in a row is a common saying, meaning that you are organized and ready for what's ahead. If you observe your ducks you will notice that they do walk in a straight line, one behind the other, most of the time – or often in a V-formation, much like ducks and geese in flight (although the walking V-formation isn't helpful for wind resistance like it is when they

fly). I believe both walking formations are related to the placement of their eyes, as mentioned above. By walking in a straight line, the lead duck only needs to keep watch to see what's ahead, which she (or he) will do by turning her head side to side as she walks, but the ducks farther down the line can keep their heads facing forward, their eyes simultaneously watching both sides for danger, and occasionally one will tilt their head to scan the sky. This works the same way in a V-formation, with the lead duck covering the front flank and the other ducks only having to worry about the sides. This way the ducks act as one mass, keeping them all safer by allowing them to cover both front and sides simultaneously.

Mud dabbling

Give your ducks a tub of water or watch them find a puddle, and they'll soon be rummaging in the resultant muddy mess, making quarter-sized holes in the ground. Believe it or not, they aren't drilling for oil or aerating your lawn. What they're actually doing is fishing for algae, worms, grubs, bug larvae and even small stones, using the serrated bristles along

edges of their bill to strain the sludge and let the water drain out while retaining the solid matter, which they love to eat. Ducks have a hard nub on the tip of their upper bills called a "nail" or "bean," which allows them to dig into the soil and scoop up food; they prefer wet soil because it makes it even easier to dig. The softer sides of the bill are used to feel around for food, much as we would use the tip of our finger.

Foot stamping

I would often notice my ducks stamping their feet up and down when it rained or when they had splashed water all over the ground. At first I thought they were just so happy to be playing in the water, they couldn't contain their exuberance, but then I read about "worm charming."

Worm charming is the method of drawing worms to the surface using vibrations that are thought to imitate rain. When it rains, worms come to the surface to avoid drowning. Obviously, if the ducks can fool the worms into thinking it's raining, and force them to the surface, the worms are easier to catch and eat.

I discovered that worm charming has been seen being used by many species of wild birds, as well as waterfowl, who all use various methods to vibrate the soil to entice worms to the surface. Tapping the ground with their feet is the most common way of worm charming, used by seagulls – and even wood turtles, who stamp their feet to attract the worms to the surface and allow the turtles to gobble them up.

In 2008, researchers from Vanderbilt University claimed that the worms come to the surface because the vibrations made by these wild birds are similar to those produced by burrowing moles, which lure earthworms in order to eat them. Either way – whether the ducks are attempting to mimic rain hitting the ground or moles digging – foot stamping is their ingenious way of tricking the worms and getting an easy meal. Fascinating as this all is, I still prefer to believe my ducks are just so happy, they can't help but dance!

Surfing in the pool

When the ducks are playing in their pool, one will often flatten her back and stretch her neck out, seeming to encourage the other ducks to use her as a sort of surf-board. Both male and female ducks will climb on top of her, first standing on her back, then lying on her and holding onto the back of her head with their bill to keep

balance. This is how ducks mate. Female-on-female (or practice mating) is normal, but it is often also used by two females as a show of dominance – or the ducks could be just "play surfing."

Mating is normally done in the water for two reasons; first, it eases the strain on the female's legs and back and lessens the chance of injury to her – and second, ducks are safer on the water than on land, so at this time when they are exceedingly vulnerable they lessen their vulnerability a bit. As part of the mating ritual, a drake will often rise up out of the water, stretch his neck out, spread his wings and flap them, while making a trumpeting sound, almost like a swan. This is the epitome of the majestic creature that is a male duck.

Post-swim preening

After they've been swimming, you will notice your ducks preening their feathers. They are able to turn their head and twist their neck to reach every part of their body and manage to get into some pretty amazing positions that I like to call "duck yoga." Ducks have a "preen gland" at the base of their tail that they stimulate after bathing to redistribute the oils onto their feathers. This is what keeps the ducks waterproofed and able to float on the surface of the water and also swim and not get waterlogged or soaked to the skin.

Nipping toes, fingers, arms or legs

Believe it or not, when a duck bites you, it hurts! Ducklings often nibble on your fingers as they try to figure out what's food and what's not. That's perfectly normal and generally once they figure out your fingers aren't edible, they'll stop. But sometimes once your ducks reach puberty, you'll have a drake (usually drakes are the ones guilty of giving "love" bites) start to grab your fingers, toes, skin on your legs or forearms, and pinch and twist. While it can be somewhat flattering – your duck is most likely flirting with you – it isn't behavior you want to allow, especially because it also means he's asserting dominance over you. You need to keep the title of alpha duck to keep the ducks in line, so you want to halt his bad behavior. When he nips you, tap him lightly on the bill with your fingertips or gently hold his bill closed and say "No!"

Standing on one leg

Flamingoes aren't the only birds who stand on one leg. Ducks also do it. They will sometimes stand on one leg to allow the muscles of one leg to rest, but more often they do it in the winter when they are on snow, ice or the cold ground. This way, they are minimizing their heat

loss. By standing on only one leg, the amount of cold blood from the foot running to their heart is cut in half, resulting in less energy expended to re-warm and recirculate the blood. They tuck the other foot into their thick, downy feathers nestled next to the body to stay warm.

The way ducks' circulation system works is this: the cooled blood that runs from their feet and legs back to the heart to warm up runs very close to the warmed blood on its way back to the extremities – so the blood arrives back at the feet already cooled, making the feet stay fairly cool. Because their feet are not much different temperature-wise in relation to the ground (or water) temperature, this greatly reduces the heat loss through their feet, since heat transfer occurs at a much greater rate when the difference in temperature between the foot and ground is more significant.

This complicated circulatory system, coupled with very little soft tissue and nerves in their feet and legs, is also what allows wild ducks to paddle around on half-frozen ponds in the winter and not get frostbitten feet. Our ducks love to sit in the snow with both legs pulled up next to their bodies, toasty warm, their bellies insulated by a thick layer of fat and several layers of feathers that trap warm air in between them.

Sleeping with one eye open

Ducks have the unique ability to literally sleep with one eye open. Their brain is split into two separate halves, with each side controlling one eye. This allows a duck to rest half their brain, while keeping the other half alert for predators. Generally, ducks will tuck their head under a wing with one eye open and watchful while they nap during the day. Ducks are fairly nocturnal and active under the cover of darkness. The have fairly poor night vision, but they do see in full color by day and also can see ultraviolet light, which gives them superior vision at both dusk and dawn when predators are

most active. They stay awake and alert at these times, so they only sleep a few hours at night, and instead prefer to nap periodically during the day.

Ducks generally sleep in a row with the ducks on either end sleeping with one eye open and on the alert, while the ducks in the middle can sleep soundly, both eyes closed, and rest both sides of their brain.

Burying eggs

Ducks have a habit of covering up their eggs after they lay them. The obvious reason is to keep them safe from predators. In the wild, when a duck gets up from sitting on her eggs

to eat, drink and take a swim, she will cover them up so a fox or other predator won't stumble upon her nest. Not only the behavior of broody ducks, our ducks pile straw on top of their eggs after they lay them as well, even though they have no intention of sitting on the eggs to hatch them. This has the added bonus of keeping the eggs warmer in the winter and not freezing before I can collect them – although it does mean that on occasion I've discovered a secret cache of eggs in a corner of the duck house.

Puffy, hissing duck

Ducks are generally happy, friendly creatures, so a hissing duck can only mean one thing: she's broody. Domestic ducks have had the broodiness bred out of them for the most part, since they are mainly raised as egg producers only (and not for hatching ducklings); a broody

duck won't continue laying eggs once she starts to sit on the nest full time. But on occasion, a duck will get into her head that she wants to sit on a nest of eggs and hatch them. She will stay on the nest and stretch her neck out and hiss at you if you try to take the eggs or remove her from the nest. She'll fan her tail feathers and puff up, trying to look larger and more foreboding. I've had a few ducks go broody for a day or so, but they seem to lose interest in no time and go back outside to play with the rest of the ducks.

Tail wagging

Unlike a grumpy broody duck's fanned tail, a happy duck will actually "wag" her tail. As near as I can tell, when a duck wags her tail it's a sign of happiness, pure and simple. Wagging duck tail, or wagging dog tail, it makes no matter – you have a happy pet!

After swimming, a duck will also shake almost like a dog, starting at the head, with the shake working its way down the body, finally to the tail, shaking the last water droplets off.

Happy quacks

Happy quacks are just what they sound like: the sounds of a contented flock of ducks proclaiming to the world how much they love life. I make it a point to bring treats to my ducks anytime I go down to the barn, even if it's just a handful of dandelion greens, and in return I am greeted with happy quacks every time!

Open bill, full-on quacking is not the norm, however; more often I find the ducks chattering among themselves. Soft quacking, with some chitter-chatter thrown in (I call it their Morse code), some whistling, squealing and whatnot is normal duck "talking." Just give them a bale of straw or a pile of leaves to play in and you'll hear what happy ducks sound like.

Male ducks, while not true quackers, will often engage in quacking competitions (they are more like raspy, honking competitions!), getting louder and louder and standing taller and taller, to show their superiority to the females.

However, if the flock's quacks become strident, rhythmic and in sync, that is the ducks' distress call. Loud, synchronized quacking by two or more ducks (and often the entire flock) could signal a predator lurking, or something flying overhead, a tarp flapping in the wind, or anything that is not normal in your ducks' world. It probably would behoove you to go take a look and see what is stressing them. Oftentimes, it is just an airplane flying overhead, but a stressed duck is not a happy duck and I will often quack quietly to my ducks when they are stressed to let them know everything is okay. I think they really do react to a calm quack – even a human one. I also quack to my hatching eggs as they begin to hatch, my own version of a mother duck pep talk.

Standing facing wind and rain

A couple of years ago, a hurricane was predicted for our area, and as we hurried to secure everything and get our horses and chickens in where it was safe and dry first – before I rounded up the ducks who don't have a problem getting wet – I happened to look into the duck pen and the ducks were all standing in place, chests puffed out, facing directly into the wind and rain. It was a very odd sight to see them all facing the same direction that way. Since then, I have noticed during downpours, especially when there's wind, that the ducks will stand still, facing into the wind.

The best I can figure is that like airplanes, ducks prefer to take off and land into the wind, so the ducks are prepared to talk off at a moment's notice, if need be, to escape a predator. (Never mind that they're in an enclosed pen and can't fly anyway, old habits die hard, I guess, and their DNA is still encoded with remnants of their wild brethrens' behavior.) Another theory I've read is that the ducks face the wind to avoid ruffling their feathers and letting the rain in underneath them. This way, their feathers lie flat and work to keep the duck waterproofed and their body dry underneath.

So there you have it. Some common duck behavior that I've observed many times with my flock. Oh, and just for the record, a bunch of ducks isn't actually called a "flock" unless they are flying. When they are on the ground they are called a "brace" or a "team." And when they are in the water they are called a "raft" or a "paddling." A bunch of ducklings being cared for by a mother duck is called a "brood."

SIX

TREATS

Ducks are omnivores. While the cliché is to feed ducks bread, that is actually extremely bad for them. In addition to being not very nutritionally well-rounded, bread can cause choking or impacted crops. Ducks love to forage, eating grass and weeds – but being true omnivores, they also eat slugs, grubs, worms, crickets, grasshoppers, spiders and all kinds of other bugs, as well as mice and small snakes, toads and lizards.

It's a community event at the feeder...

🦆 Duck Treats vs. Chicken Treats 🦆

Ducks can eat the same types of treats that chickens do. In fact, since our chickens and ducks share the run, they are all offered the same treats – but our ducks are far more partial to leafy greens and vegetables than are the chickens. It's interesting to give them a selection of foods and watch which they gravitate to first. That insight will give you an idea of your ducks' favorites. Our ducks' favorites are spaghetti, watermelon, corn and mealworms, in addition to anything green.

The treats I feed my ducks vary a bit depending on the season. I try to grow as much as I can for the ducks and also take advantage of the local flora to keep things economical – and to provide them as natural a diet as I can, mimicking what wild ducks might eat.

Treats should be limited to healthy foods and to no more than 10% of your duck's total diet; however, weeds, herbs and grass are considered "green" treats and can be fed in unlimited amounts.

Ducks don't have teeth, although they do have serrated bills, but to make things easier for them I usually cook, puree or grate things that are tough, like broccoli stalks.

Early Spring/Spring

Chickweed and dandelion greens are at their peak in early spring in the more temperate climates. Our ducks love both and they are a great diet supplement.

Other early garden crops such as fresh peas are a favorite treat. They especially love peas floated in their water tubs, but they enjoy snap peas in the pod as well. And they are fans of early lettuce. Ducks will generally only eat greens if they are fresh. Wilted, trampled greens will be ignored, so floating greens in their water tub or swimming pool keeps the greens fresher and less goes to waste. Anything green floating in the water will make ducks happy.

Fresh, early spring greens and peas are irresistible, especially when they're floating in water.

Year-round, ducks love to "bob" for treats in their water tub. Grow extra crops of kale, cabbage, collards, chard, lettuce, Brussels sprouts, beet greens and spinach for your ducks to supplement their diet.

Dill from the herb garden makes a great treat, as it is an antioxidant, relaxant and supports healthy respiratory health. Parsley, another early herb to show up in the garden, is high in vitamins and is a laying stimulant.

Feeding too much spinach can interfere with calcium absorption and reduce the amount absorbed by a duck's body, thereby causing egg binding in females, or soft-shelled eggs. Spinach is extremely nutritious but should be fed in limited amounts only. The effects of the oxalic acid can be partially mitigated by adding some apple cider vinegar to the water any time you serve spinach. **Note:** *other foods that contain oxalic acid include beet greens, collards, kale, Swiss chard and turnip, so adding apple cider vinegar to your ducks' water any time you serve these treats is a good idea.*

Summer

Cut grass is a nutritious treat for your ducks and provides a healthy dose of protein, as well as magnesium, iron and phosphorus, among other nutrients. But only cut untreated grass for them – no pesticides, fertilizers or herbicides – and cut it into 1" or shorter lengths to avoid impacted crops. I look for tender, young grass because it's less fibrous and less likely to get caught in the ducks' gizzards.

Water-laden treats such as watermelon and cucumbers are beneficial and help keep your ducks hydrated during the hot summer months. If you can grow them in your garden, all the better.

Berries are big summer favorites and include strawberries and tops, blueberries, raspberries and blackberries. Other popular fruit treats: pears, and pitted cherries, plums and peaches. *Note:* I remove the pits of any stone fruit I give our ducks because ducks are notorious for trying to eat everything.

Watermelon will help keep your ducks hydrated in hot summer months.

Don't feed ducks unripe tomatoes or their leaves, vines or stems, since they can be toxic.

Smartweed (aka lady's thumb or heart's-ease) has many health benefits for ducks and humans alike.

Tomatoes…and another caution. Ripe tomatoes are a favorite treat on occasion, but avoid feeding your ducks green tomatoes, or the plant's leaves, vines or stems, all of which can be toxic. (Tomatoes, eggplant, rhubarb and white potatoes are all in the nightshade family and should be avoided. The ripened fruit contains less toxin than unripe fruit or other parts of the plants, but regardless, should be fed only in moderation. Err on the side of caution.)

More summer garden treats include corn, broccoli, zucchini and summer squash.

Herbs. My herb garden is in full swing all summer and the ducks get large helpings of the various culinary herbs including basil, cilantro, mint, oregano, parsley and sage. Mint in particular is beneficial in the summer for its natural cooling properties. (See the Appendix for Edible Herbs, Weeds & Flowers, page 136.)

Smartweed, also called lady's thumb or heart's-ease, is a perennial herb that grows in abundance in the summer and is another duck favorite. It has various health benefits including being an anti-inflammatory, anti-oxidant and good for respiratory system support. Smartweed also combats bacterial infections and guards against E. coli and staph infection – so the ducks benefit from having smartweed added to their diet.

Fall

With cooler temperatures comes the return of the chickweed and dandelion greens, so the ducks fill up on those, along with the spent vines and any remaining vegetables from the garden.

Squash and pumpkins from the garden: a real duck favorite. The guts and seeds can be fed raw, but I like to steam, bake or boil the flesh to make it easier for the ducks to eat. Seeds of squash and melon are thought to have a coating that works as a natural wormer, so feeding your ducks plenty of them is very beneficial.

Sweet potatoes, skins and all, are a nutritious treat, either raw or cooked – but again, it's easier for the ducks if you cook them. Sweet potatoes are part of the morning glory family and safe to feed ducks. However, white potatoes (as noted above) are in the night-shade family and all parts of the potato and plant contain the potentially harmful toxin. I err on the side of caution and only feed our ducks sweet potatoes, which are more nutritious anyway.

Yum...grubs!

Since molting season is upon them, extra protein is a good idea to help them grow in their new feathers, which are almost entirely made up of protein – so I offer dried mealworms on a regular basis and the ducks really love them. Some other good fall treats: insects, earthworms, grubs, meat scraps and oatmeal (either raw oats or cooked).

PLANT-BASED SOURCES OF PROTEIN

Alfalfa	Mung Bean Sprouts	Sea Kelp
Broccoli	Oats (raw or cooked)	Spinach
Cauliflower	Pumpkin Seeds (unsalted)	Spirulina
Lentils (cooked)	Quinoa	

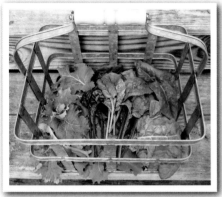
Spinach is rich in protein as well as Vitamins A and K.

Spirulina has a whopping 64 grams of protein per cup, making it a protein powerhouse. Cooked lentils are also jam-packed, with 18 protein per cup, while quinoa provides 8 grams per cup. Cauliflower also provides omega-3s and niacin (niacin, which helps with bone growth, can be found in broccoli and oats as well). Pumpkin seeds work as a natural wormer and contain 9 grams of protein per ounce. Spinach contains protein, but also Vitamin A (for the immune system) and Vitamin K (helps blood clotting and bone growth). All these foods are most beneficial in the fall, but make for healthy treats year-round for your flock.

HERBS WHILE THEY'RE MOLTING

Basil	Dill	Parsley
Celery Seed	Fennel	Saffron
Chervil	Marjoram	Spearmint
Coriander	Oregano	Tarragon

These twelve herbs are high in protein, so adding them fresh or dried to your ducks' diet is beneficial during the molt.

Grains such as wheat and barley, as well as grasses, contain protein, so a free-ranging flock will find sources of natural plant protein quite readily.

Of course meat scraps are also good protein sources. Cooked beef, chicken, turkey and pork, and cooked fish (skin and all), are great for molting ducks. Your ducks will also love cooked shrimp (shells and all), feeder fish, minnows and slugs.

Winter

Treats that help keep the ducks warm are in order once winter sets in. Cracked corn is a common treat; also, unsalted raw peanuts in the shell. If you are worried about choking, you can grind the peanuts up into small pieces to feed your ducks. The energy they use to digest the peanuts helps warm up their bodies. Corn isn't terribly nutritious but it is high in calories and carbs, so it provides good energy in the winter.

On cold mornings, warm oatmeal will be appreciated by your ducks. Whole grains are more nutritious than white. Other good choices for treats: cooked brown rice, whole wheat pasta, non-sugar cereals, cooked quinoa, millet and barley. On occasion stale whole wheat bread is okay, but avoid giving white bread, fried foods or salty crackers. Ducks gain weight easily and added weight puts a strain on their legs and feet.

Scrambled eggs are another nutritious treat your ducks will love.

I like to sprout wheat grass, mung beans and other grains for our ducks to enjoy through the winter. And they love wheat berry fodder grown indoors.

Cracked corn provides extra energy for wintertime. Unsalted raw peanuts, too.

Winter treats are heavy on the whole and sprouted grains.

Better safe than sorry checklist

Limit these:

- Asparagus – can taint the taste of eggs
- Citrus – can lead to soft-shelled eggs
- Dairy – can cause diarrhea
- Iceberg Lettuce – can cause diarrhea
- Mangoes – can cause itchy mouths in your ducks like they do for some humans
- Spinach – can lead to soft-shelled eggs
- White rice, bread, pasta – low in nutrition

Avoid these:

- Anything moldy or spoiled
- Apple and cherry seeds – toxic, so avoid those altogether; peach and apricot pits, as well.
- Avocados – contain toxin
- Chocolate, coffee grounds, tea bags, alcohol
- Nightshade family, especially vines, leaves and skins – rhubarb, green tomatoes eggplant and white potato
- Nuts and large seeds – don't digest well and can cause choking or get stuck in the crop since ducks swallow their food whole. If you do feed nuts or seeds to your ducks, they should be ground first.
- Onions
- Raw dried beans
- Salty, sugar or fried foods

SEASONAL TREAT CHART

SPRING

Nourishing treats to prepare for laying/hatching season

Basil
Chickweed
Cilantro
Collards
Dandelion
Dill
Kale
Lettuce
Parsley

SUMMER

Hydrating and cooling treats to battle summer heat

Corn on the cob
Cucumber
Mint
Tomatoes
Watermelon

FALL

Protein-rich treats to help grow in new feathers

Broccoli
Mealworms
Meat scraps
Mung bean sprouts
Pumpkin
Spinach
Squash

WINTER

Energizing and warming treats to battle the cold

Cracked corn
Oats
Scratch grains
Unsalted peanuts
Unsalted popcorn

Natural Sources of Niacin, Calcium and Protein

A well-rounded, balanced diet is as important for ducks as it is for humans. Three nutrients in particular play a very important part in the diet of a duck: niacin, calcium and protein. Focusing on healthy treats that provide these nutrients to your ducks will help you raise a healthy flock. And the best part is...our ducks love all of these foods! (Of course, your duck's main food should always be a good quality layer feed or waterfowl feed.)

Niacin (Vitamin B3)

Ducks need more niacin than chickens for strong legs and proper bone development. Niacin also helps with heart health, improving blood circulation and brain function. Each duckling can need slightly different amounts of niacin, so adding some natural sources to their diet, in addition to adding brewer's yeast to their feed, can be beneficial to growing ducklings.

Grown ducks don't generally need any more niacin than is included in regular chicken layer feed, but adding these same natural sources of niacin can ensure they are getting enough.

Good sources of niacin include: barley, cooked beans, brown rice, cantaloupes, corn, cooked eggs, cooked lentils, peanuts, peas, pumpkin, salmon, sardines, squash, sweet potato, tomatoes and tuna. Leafy greens such as beet, mustard and turnip greens, collards, kale and Swiss chard contain lesser amounts of niacin.

Niacin for can be found in many grains, vegetables, fish and leafy greens.

Calcium

Calcium is added to layer feeds to help with strong eggshells, strong bones and the muscle contractions necessary for your ducks to lay their eggs. It's also recommended to supplement that calcium with free-choice crushed eggshells or oyster shells.

You can supply your ducks with other nutritious sources of calcium, which include: alfalfa, burdock root, chamomile, chickweed, clover, dandelion greens, horsetail, lambsquarters, mustard greens, nettle, parsley, peppermint, raspberry leaf, rose hips and watercress.

Protein

Adequate protein is essential to your ducks' diet for optimal overall health and especially during the molting season for feather re-growth. I do feed my ducks meat scraps such as cooked fish, steak or even poultry. Left to their own devices, ducks will eat not only worms, bugs and grubs but also frogs, toads, small snakes and lizards – all excellent sources of protein. But there are plenty of plant-based protein sources that are a healthy addition to their diet if you don't feel comfortable feeding your ducks meat, or don't eat meat yourself or as a family.

Plant-based protein sources include: alfalfa, broccoli, cauliflower, kelp, lentils (cooked), mung bean sprouts, oats (raw or cooked), quinoa, spinach and spirulina.

Fresh, nutritionally rich treats will entertain your ducks and contribute greatly to the health of your flock.

SEVEN

DUCK HOUSES AND PENS

Since domestic ducks are so vulnerable to predators, they need to be locked up at night most definitely, and also in a safe pen by day if you aren't home and able to keep an eye on them. As long as they are out of the wind and inclement weather, their shelter can consist of a simple structure with some straw on the floor and a securely locked door. Ducks don't roost or perch and are perfectly content making a nest on the ground in the straw on which to sleep. They don't generally use nesting boxes, but will tend to lay their eggs in a corner of their house in the straw.

At night

Ducks, being extremely cold-hardy, are perfectly happy being outside all day in the rain or snow and you may be tempted to leave them outside in a pen all night, which they would certainly enjoy! One caveat though: it's very hard to build a pen that is 100% predator-proof – and under the cover of darkness, a clever raccoon can open latches, rip through or reach through fencing, and climb. Dog, foxes and wolves can dig under fencing and weasels can slip through.

Keep your ducks safe inside a pen.

Reinforce coop doors with locks and windows with wire to protect against predators.

By far the safest for your ducks is a coop or house with ½" welded wire on all the windows and vents, and a locking eye hook latch with a carabiner or other predator-proof lock on the door. Allow for at least four square feet of floor space inside the house per duck, giving them enough room to stretch their legs and flap their wings. Ducks emit lots of moisture when they sleep at night, so any housing needs adequate ventilation, summer and winter, which means lots of open vents positioned up high, year-round.

My duck house is situated inside our large, covered run. I have a night pen attached to the duck house. It's a small 4x8 enclosure made with a wooden frame and 1/2" welded wire; it is completely enclosed within the larger run, This added level of security allows my ducks to choose to sleep in their house, or in the attached pen under the stars (which is generally what they prefer), while being protected from predators behind two sets of fencing. This also allows me to leave them feed and water overnight outside in the pen, so as not to attract flies or moisture inside their house. I highly recommend this setup for the ducks. This keeps them safe, keeps their house bedding dry and is healthier for them, since they have access to the outdoors and fresh air all night if they wish.

Housing your ducks with your chickens

I have never read a specific temperature that is too cold for cold-hardy ducks, but from my own observations, on any night that is predicted to go below 16 degrees F in the winter, the ducks march into the chicken coop on their own and settle into the straw on the floor in the coop. Somehow I guess they think they will be warmer in the coop than in the duck house on those abnormally cold nights, or maybe they know their body heat will help keep the chickens warmer! Regardless, I let them decide and sure enough, if the next night is going to stay above 16 degrees, the ducks are back in their own house that night.

You can certainly house your ducks with your chickens on a permanent basis, and they will sleep and lay their eggs in the corners of the coop in the straw; just be aware of the high moisture levels and make it a point to change out the dirty litter regularly. You will also be better off not leaving feed or water in the coop at night if your ducks share the coop. Your coop floor will just end up being a wet mess that attracts flies, rodents and predators.

Ducks and chickens can share the coop, but the ducks can be messy roommates!

75

Ducks do tend to attract flies, since they keep the feed moist and wetter than chickens do. I hang bundles of fresh herbs that are known for their fly-repelling properties, including basil, dill, lavender, lemongrass, mint and rosemary.

I also make herbal fly spray to use in the duck house and around the feed. It's all-natural and won't hurt the ducks one bit. You can experiment with different combinations, but I like the Vanilla Mint best.

Vanilla Mint Fly Spray

This spray has a white vinegar base, which has the added benefit of being an antibacterial and disinfectant. I mix the citrus scent of a lime with basil and mint, all of which are particularly unappealing to flies. I throw in a vanilla bean for good measure, since vanilla seems to be an excellent fly repellent as well.

What you'll need:

1 lime
Fresh mint leaves
Fresh basil leaves
Vanilla bean, split

White vinegar
Pint Mason jar
Spray bottle

What to do:

Cut the lime into slices or quarters and place in the canning jar along with a handful of fresh-cut mint and basil leaves, torn slightly with your fingers to release their oils. Add the split vanilla bean, then fill the jar with white vinegar leaving at least ¼" headroom. Set the jar in your pantry, cupboard or on the kitchen counter to "age" for about two weeks, shaking every few days to reinvigorate the contents. When ready to use, strain the liquid into a squirt bottle and spray liberally wherever you see flies congregating – in the coop, duck house, around the feeders or run. Repeat as needed.

Shelter from the wind

While ducks don't mind cold temperatures or precipitation, they don't seem to like wind much and in the summer need shelter from the sun. They should be happy being outside in their pen year-round, especially if you set up some sun and wind blocks for them. Stacked bales of straw work great, or a tarp or sheet of plastic wrapped around one corner of the pen, or some simple wooden A-frame structures they can seek shelter under as needed – all will be appreciated by your ducks. For a natural wind and sun barrier in the warm months, you can grow squash, peas, grapes or other vines up the sides of your duck pen. Then your ducks will have the added fun of munching on all the leaves and fruits they can reach!

There aren't many duck houses manufactured commercially, but it's easy to build one yourself or repurpose a dog house or playhouse into a duck house. These cute ideas might get you inspired to see what you can use for your duck house.

A duck shelter can take many forms.

EIGHT

DUCK POOLS

It's a common misconception that domestic ducks need a pond or other large body of water to be happy and stay healthy. They will be perfectly ecstatic with a plastic kiddie pool, horse trough or tub large enough to play in. You need only watch ducks in the water to realize the sheer bliss they experience playing, splashing, preening or just floating around.

One of the distinctive features of domestic ducks is their webbed feet, which allow them to paddle and easily swim in the water. When ducks push their feet back in a kicking motion, the webbing catches the water and propels the duck forward. The webbed feet create the peculiar waddling motion when ducks walk on land, making them slow and awkward out of the water.

Nothing gives a duck more joy than time in the pool. Sheer bliss.

This deep tub is set up with steps for easy entry and exit.

In addition to providing a water tub deep enough for your ducks to completely submerge their heads, ensuring they can clean their eyes, nostrils and sinuses, you should also provide them a place to bathe at least several times a week, if not daily – a kiddie pool or a larger tub. They use the water not only to clean off their feathers, but as they preen after their swim they stimulate their preen gland, which helps distribute the waterproofing oils over their feathers. This keeps them protected against the elements. Ducks have extremely waterproof feathers due to an intricate feather structure and a waxy coating that is spread on each feather while preening. Even when the duck dives underwater, its underlayer of down will stay completely dry.

If the pool or tub is fairly deep, be sure to set up "steps" both inside and outside so the ducks can easily enter and exit the pool. Stone pavers, bricks or cement blocks all work just fine. And remember that ducklings hatched in an incubator should never be allowed unsupervised swims until they are at least a month old because they can be chilled and die of hypothermia or even drown, since they are not waterproofed yet. Even if you have ducklings that were hatched under a duck and have had maternal guidance, be sure to supervise the swims while they are still small, so older ducks don't accidently drown the little ones.

A caution about ponds: If you do have a pond on your property, your ducks will love to be allowed access to it, but be aware that foxes, dogs, weasels or minks might be lurking, snapping turtles or alligators sometimes bite off ducks' feet, and the ducks are easy prey for owls and other aerial raptors while in the water. Also, you might have trouble getting them to leave the pond at dusk so you can lock them up safely for the night.

...and swimming pools: Ducks should never be allowed to swim in your family swimming pool. The chlorine and chemicals are not good for them and can strip the oils from their feathers, and you certainly don't want your family swimming in water in which ducks have defecated.

Whether it's a natural pond or man-made pool, ducks love their water!

Three is just right for this kiddie pool.

What kind of pool works best? As far as ducks and water goes, any container will be considered fair game to them. I have even caught our ducks sitting in our dog's water bowl! They will swim in the horse trough, in mud puddles, even in water buckets.

You can get fancy and find an old garden tub or bathtub on Craig's List, on the curb on trash pickup day or at a Habitat for Humanity store; sink it into the ground and plumb it to drain outside of your duck pen. You can build a wooden "sundeck" complete with steps for your ducks to use to get into their kiddie pool (or just stack some cement blocks or pavers to help your ducks get in and out). You can buy a koi pond kit, with filter, and install that in your duck pen – or simply dig a hole in the ground, line it with heavy duty plastic and fill it with water. Plain or fancy, your ducks don't care as long as they have a place to swim safely. Whatever you decide, it has to be easy to drain, clean and refill, as well as easy for the ducks to get in and out of.

Mating behavior. Ducks generally mate in the water. This helps to alleviate any strain on the female duck's legs and feet and also provides the ducks more protection from predators when they are at their most vulnerable. The female duck will flatten her back to allow the drake to climb atop her. He will grasp her back with his feet and grab hold of the back of her head with his bill, both to help keep his balance and to hold her head up out of the water.

Clear signs of over-mating in a female are bare patches at the back of her head and under her wings. If you see evidence of over-mating, you might want to add more female ducks to your flock, separate the female who seems to be a favorite, or separate the males from the flock for a while to let all the females recover. Another sign of over-mating is bubbling or runny eyes. This is a symptom of an eye infection caused by bacteria in the drake's saliva. (See page 95 for more about this.)

Fresh, clean water. Remember that the pool water should always be fresh, because algae or bacteria can be quite harmful to your ducks, but muddy water doesn't necessarily always equate to dirty water. Ducks manage to muddy up their swimming sources faster than you can dump and refill them, and if you strive for crystal clear water you will just make yourself crazy. Instead, just be sure the water is dumped out, the containers are scrubbed clean (I use a white vinegar/water mixture) and refilled at least once daily in the warm months and every few days in the winter. Mud and dirt are not a problem in the water, but you don't want to leave water that has feces in it or starts growing algae.

Mating behavior can be "real" or "play."

I refill and clean the pools once daily in warm months and every few days in the winter.

NINE

A CLEAN BILL OF HEALTH

Ducks are generally very hardy animals, not particularly susceptible to many of the health issues that can plague chickens, which can include Marek's disease, Newcastle disease, parasites or coccidiosis. A duck's normal body temperature hovers around 106-107 degrees F. This feverishly high temperature means that pathogens and bacteria don't survive well inside a duck's body. Additionally, due to all the time ducks spend in the water, they aren't usually bothered by scaly leg mites, fowl mites or lice, as long as they have regular access to clean water for bathing. In fact, providing plenty of clean water for your ducks to drink and swim in can keep most health issues, including eye and sinus problems, at bay. Ducks tend to stay extremely healthy when kept in an uncrowded, clean environment, provided fresh water and fed a healthy diet. But sometimes, despite our best intentions, things do go wrong.

Most duck health issues revolve around three things:

- The ingestion of toxins or foreign objects – since ducks tend to try to eat anything they can. Things like rat poisons, fly sprays, common household chemicals and even stagnant water, can quickly cause severe distress or death in backyard ducks.
- Injuries from predators – since domestic ducks can't fly and with their webbed feet are extremely slow and awkward on the ground.

- Foot and leg injuries – especially in the heavier breeds and especially if you let your ducks get overweight or don't provide water for them in which to mate.

A natural approach. I don't advocate using antibiotics or commercial medications with any of our animals, and our ducks are no different. Being fairly hardy, they really should not need much medical care anyway, and most of the common duck ailments can be treated naturally. At the first sign of anything wrong, some healthful additives can help bolster your ducks' immune systems and build good bacteria. If you aren't already doing these things on a regular basis, I highly recommend the following to maintain optimal health:

A natural approach to duck health can prevent problems down the road.

- Add a fresh garlic clove and one tablespoon of apple cider vinegar to each gallon of water to help boost the immune system, several times a week. Alternatively, feed fresh minced garlic free-choice.

- Add probiotic powder to the daily feed to boost good bacteria and help fight infection.

- In addition, offer a "tea" of steeped echinacea root, dandelion leaves/roots, parsley and cilantro or offer fresh thyme, oregano and sage free-choice on occasion to bolster immune systems.

Of course, the earlier you catch anything wrong, the easier it will be to treat it. Handling your ducks often is so important to check for clear eyes, regular breathing, good body weight and feet free of any cuts. Make it a point to give each duck a quick "checkup" on a regular basis to head off any potential problems before they develop and worsen. *Note:* See the Duck First-Aid Kit on page 108.

The following are some of the more common health issues you might encounter in your backyard flock:

Poisoning/Botulism. Especially if your ducks roam free in your yard, there is a chance they might ingest something they shouldn't. Rat poison, antifreeze, fertilizers, mothballs, insecticides, etc., can all be toxic to ducks. Ducks also love to eat shiny things, including screws, nails, nuts, bolts, wire, staples, glass, spare change or pieces of lead, zinc or other metal – which can lead to poisoning or internal injury. Botulism is also a concern and contracted from bacteria found in stagnant water or decaying animal carcasses. If your duck drinks water from contaminated ponds or puddles or eats spoiled or rotten food, she can contract botulism. Symptoms usually include respiratory distress, diarrhea, twitching, limpness, and loss of muscle control – specifically the inability to hold her head up – with death occurring within 24-48 hours if not treated. Symptoms of poisoning include lethargy, decreased appetite or weight, difficulty walking, seizures and possibly death if not treated.

What to do: If a visit to a vet is not possible, add 3 tablespoons of blackstrap molasses to a gallon of water and offer that as the sole liquid for 8 hours, then offer plenty of clean, fresh water. The molasses can be effective in flushing the toxins. Charcoal is also an effective antitoxin and the contents of charcoal pills can be sprinkled over your ducks' feed if they are still eating.

If you believe your duck swallowed something sharp, a visit to the vet for an x-ray and possible surgery is highly recommended.

Prevention is far easier than treatment, so you should be super vigilant about checking your yard and run for odds and ends the ducks might ingest and preventing areas with standing water.

Predator Attack. Sadly, predator attacks are very common, since domestic ducks can't fly and aren't able to get away quickly. Foxes, dogs, raccoon, possums, hawks, owls and coyotes are all a threat. If your ducks are attacked, don't despair, because ducks can be resilient – but they do need some immediate care to survive, depending on the severity of the wounds. Bring the injured duck inside and keep her calm – in the bathtub with some towels is perfect. (And be sure the rest of your ducks are safely locked up, because the predator will be back.)

Ducks are vulnerable to predators. Immediate first-aid is key.

To treat bites, torn skin, or claw injuries: Clean your duck up with warm water as best you can. Some cornstarch or flour can stop minor bleeding; more serious wounds might require pressure. Once the wounds are clean, smear them with honey or Green Goo. Don't try to remove any broken feathers or torn skin. Leave it to heal. Smear with honey any cuts or scrapes on the bill, legs or feet. (Eye injuries should be treated as below under "Eye Issues." Leg or foot injuries should be treated as below under "Leg or Foot Injuries.")

Once you have your duck cleaned up and patched up, offer her some sugar water, Nutri-drench or electrolytes such as plain Pedialyte. Dip her bill in the water, but *don't* force her to drink with a syringe or eyedropper. Moisten some feed for her, especially if she has bill injuries. Don't force feed, but try offering favorite treats if she's not eating in a day or so. Chickweed, comfrey and calendula can help with pain relief.

Keep her quiet and indoors until she heals. Some lavender, lemon balm or chamomile can help your duck relax and soothe her. If the wounds appear life-threatening, not improving or seem infected, a visit to a vet is in order.

Leg or Foot Injuries/Limping/Inflammation. Domestic ducks are prone to injuring their legs and feet since they really weren't designed to be walking on the ground all the time. Wild

A neoprene bootie keeps a recovering foot dry and clean.

ducks can fly and spend much of their time floating around in the water, not walking on hard land. They are also smaller. The larger, heavier domestic breeds can easily sprain or injure their legs after tripping or stumbling over something (ducks are notorious for not looking where they are walking!) or tear toenails or the webbing on their feet. Torn toenails will bleed. Some cornstarch patted to the nail bed will help stop the bleeding. Clip the broken toenail if you can manage it without further injury to the duck and then wrap in gauze and Vetrap to heal. Torn webbing won't grow back, so there's not a lot to do in that case.

Domestic ducks often don't have water accessible in which to mate, so mating on the ground can lead to leg injuries in female ducks, as can entering and exiting the pool. As described in Chapter 8, ducks usually mate in the water, which takes the weight off the female duck's legs.

If you notice one of your ducks limping badly with a swelled foot or leg, or any bruising, consider separating her if that won't cause unnecessary stress (ducks get so attached to the other flock members that often separating one is worse than leaving her in with the flock.) I find that a minor limp will usually clear up on its own in a few days, but if the duck is unable to move well, or stand at all, you should separate her to avoid trampling; and be sure the duck is able to reach food and water. A small pen or cage inside the duck run works well, so the injured duck isn't apart from her friends but is kept relatively immobile and off her foot.

A limping duck should be allowed time to swim every day. The swimming will help her heal and take weight off her feet. After the swim, you can do several things to help alleviate her inflammation and pain.

These herbs can help with swelling and reduce inflammation: calendula, chickweed, plantain, comfrey. Also, apple cider vinegar.

Treatment options:

Chickweed can help reduce inflammation.

- Press fresh chickweed or plantain against the injured leg and hold in place with Vetrap, or other non-adhesive tape.

- You can also steep any of the herbs in hot water and then apply the cooled compress to the leg, repeating several times a day with a new application.

- Warmed apple cider vinegar applied topically can reduce swelling.

- Chickweed or calendula taken internally is a natural pain reliever and ducks love it, so offering your patient a big bowl of fresh picked chickweed can help them feel better in more ways than one, or you can steep them into a tea for your duck to drink.

Ducklings can develop weak legs if they are not fed a diet rich in Vitamins B and D. You can ensure healthy bone formation and leg strength by adding brewer's yeast to your duckling's diet and allowing them time outdoors in the sunlight on fresh grass. Adding brewer's yeast to an injured duck's diet, if you don't already, can also be extremely beneficial.

If you suspect a broken bone, please consider a visit to your vet. Broken bones require professional treatment, but once set, adding some specific nutrients to your duck's diet can help the bones heal faster. (See "Sources of Essential Nutrients" on page 107.)

Less common duck health issues you should never encounter, but it doesn't hurt to be aware of:

Angel wing can be corrected if caught early.

Angel Wing/Dropped Wing. Angel wing is an affliction most often seen in ducklings under 16 weeks old, caused by a too-high protein diet that causes them to grow too fast. As their bones and wings are growing and their wing feathers become heavy with blood, the wing begins to turn out from the duck's body, hence the name angel wing. It is usually correctable if caught early on, by taping or binding the duck's wings to its body for a week or two with a non-adhesive flexible bandaging tape such as Vetrap – at the same time adding oats to their feed, up to a 25% oat/75% feed ratio, to reduce the protein levels. Dropped wing is a similar affliction, which causes the blood-heavy wings to drop when the duckling doesn't have the strength to hold them up. You may notice the wings drooping and the duckling repeatedly pulling them up toward her body before they drop again. Taping and diet change can help with dropped wing as well. Once the feathers are fully formed, the blood dries up in the quills and the feathers (and therefore wing) become lighter and more manageable for the duckling.

Allowing your ducklings access to the outside to free-range and offering a variety of leafy greens and other nutritious foods can also help prevent and correct it, as can plenty of swim time for your growing ducklings Although disastrous to wild ducks who need to fly, in domestic ducks angel wing is merely an aesthetic issue, and a duck can live quite comfortably with the condition.

Ascites (Water Belly). Ascites syndrome, also known as "water belly" is a non-curable condition in which the liver and blood vessels swell and cause fluid to leak into the abdomen, causing it to expand and feel squishy like a water balloon.

Although the condition can be genetic, it is often caused by a poor diet that includes excess sodium and leads to hypertension. Ducks that grow too fast or experience chilling during their first three weeks of life have a greater probability of suffering from water belly. The duck can appear normal otherwise, and continue eating and drinking. A vet can drain the fluid to give your duck some relief, but that is only a temporary fit and it will likely recur.

Feeding your ducks a healthy diet – low in salt – and allowing them plenty of room to exercise is the best way to prevent water belly, along with maintaining the proper temperature in your brooder for your ducklings.

Aspergillosis/Aflatoxin. Ducks can suffer respiratory distress from inhaling fungus and mold spores that develop in wet bedding or spoiled feed. Gasping, labored breathing, listlessness or dehydration can all signal aspergillosis. Advanced cases of aspergillosis can lead to reduced growth in your ducks, organ failure or even death.

Aspergillosis can easily be prevented by keeping litter and feed fresh and dry. Watch for moldy feed caused by your ducks' habit of going back and forth from their water to their feed. Any feed leftover at the end of each day should be disposed of.

To prevent aspergillosis, litter and feed need to be kept fresh and dry.

Hay should not be used as bedding because it is too "green" and has the tendency to mildew quickly if it gets wet or is exposed to moisture. Straw is dry and less likely to cause problems, but should always be checked carefully before using in your duck house for signs of any mold. Removing the offending litter as soon as you become aware of the mold should alleviate symptoms. I don't leave feed or water in my duck house for this reason. It's very hard to keep the litter dry if you are putting water in the duck house.

Bumblefoot/Staphylococcus Infection. More common in the heavier breeds, bumblefoot is a staph infection caused by a cut or splinter on the bottom of the foot that allows the bacteria to enter the body and eventually the bloodstream. Bumblefoot can be easily identified by the telltale black scab on the bottom of the foot, and is often accompanied by limping, swelling or a leg that is very warm to the touch. It is far easier to treat the earlier it is caught, so doing a regular check of your ducks' feet should be part of your routine. At least every month or so, I try to check each duck one by one.

Typical bumblefoot black scab (r.).

Neoprene booties allow an infected foot to safely heal.

If you see the black scab or any callus on the bottom of the foot, soak the foot in warm water with Epsom salts to soften the skin, then carefully remove the scab with tweezers or carefully scrape it off with a sharp scalpel. Then smear the area with honey, some antiseptic salve such as Green Goo or my homemade bumblefoot salve (recipe below) to help draw out the infection and prevent it from worsening. Wrap the foot in gauze and Vetrap. Repeat the dressing at least twice a day.

If the infection is further along or the salve doesn't seem to be helping after several days of treatment, then removing the infected "kernel" might be necessary. This entails soaking the foot in warm water with Epsom salts, then making a cut in the scab with a sharp scalpel and extracting all the white stringy matter and hard kernel with a pair of tweezers. (Be sure to have some cornstarch or flour on hand to stop the bleeding.) I recommend an application of salve and dressing and then putting a neoprene bootie on the affected foot to keep it dry and clean.

Whether you are treating only with the salve or you have extracted the infected mass, keep treating the foot with the salve until a new scab forms. As long as it's not black, you've successfully removed the infection.

Herbal Bumblefoot Salve

What you'll need:

½ cup coconut oil
1-2 sprigs fresh lavender
1 bunch fresh sage
5-6 sprigs fresh thyme
1-2 sprigs fresh rosemary
1 ounce beeswax pastilles
⅛ teaspoon liquid vitamin E
⅛ teaspoon powdered turmeric
⅛ teaspoon cinnamon

What to do:

Add the herbs to the coconut oil over low heat in a double boiler, tearing the leaves a bit to release the essential oils. Steep for 45-60 minutes over low heat, then strain out the herbs. Add the beeswax to the infused oil. Return to the heat until the beeswax is melted. Remove from heat and stir in the Vitamin E, turmeric and cinnamon until well mixed. Pour into a small covered metal or glass container and allow to cool and set.

Working to prevent bumblefoot is the best option. Keep the run or area your ducks have access to clear of any sharps sticks, stones, wire, pinecones or other things that could cut your ducks' feet; this can help prevent it – as can ensuring your ducks are getting plenty of Vitamin H (biotin).

Biotin and Vitamin A: It's thought that bumblefoot is partially caused by a diet low in biotin. Good sources of the vitamin that your ducks will love include brewer's yeast, broccoli, cabbage, cucumbers, eggs, whole grains, kale, oats, raspberries and Swiss chard. Additional Vitamin A can also help prevent cases. Good sources of Vitamin A your ducks will love include carrots, cantaloupe, leafy greens, pumpkins and sweet potatoes.

Coccidiosis. Ducklings are not nearly as susceptible to coccidiosis as are chickens. Coccidiosis is an intestinal parasite often picked up in overcrowded, dirty brooder conditions. Nearly every poultry run contains some level of coccidia, but generally healthy birds develop a natural resistance over time. Adding apple cider vinegar to your ducklings' water from hatch through their life can help combat coccidia as well as other internal worms (1 tablespoon per gallon of water for maintenance several times a week)

Diarrhea/Digestive Issues. Duck fecal matter is extremely watery, especially in the summer when they are drinking more water, and can range in color from dark brown to light tan to bright green and everything in between, depending on what they have been eating (for instance, feeding an excess of beet greens will result in bright teal poop – don't ask me how I know that!) However, if you suspect one of your ducks is suffering digestive distress (example: has a poopy rear end), offering a bit of plain yogurt, brewer's yeast, fresh dill, grated apple, or cooked brown rice to her can help clear it up. Adding probiotic powder to your ducks' daily layer feed (in a 2% ratio) should help keep their digestive tracts healthy and the bad bacteria that causes diarrhea in check. Since diarrhea depletes the body of essential nutrients as well as liquid, offering some vitamin-rich plain Pedialyte or Vitamins & Electrolytes is also beneficial.

Egg Binding. Ducks don't generally suffer from egg binding (the inability to completely expel an egg) nearly as frequently as chickens do, possibly because they spend so much time in the

Ducks need calcium in their diet to help prevent egg binding.

water (soaking an egg-bound hen in a tub of warm water is the common treatment). However, it can occasionally happen due to genetics that cause her to lay extremely large or double yolk eggs, or due to weight issues or insufficient calcium. Calcium is needed for the proper muscle contraction required to lay eggs. If your duck is pumping her tail end or seems to be straining, waddles like a penguin, or is panting, she could be egg bound. 1 cc of liquid calcium taken internally by eyedropper can help, as can a warm soak. You want to take care not the break the egg stuck inside her, so if you can catch your duck without too much trouble, gently carry her inside or to a quiet place where you can fill a large container with water.

Carefully set your duck in a tub of warm water and Epsom salts for about 20 minutes, then lubricate around her vent with olive oil or vegetable oil and put her in a quiet, dark spot, maybe in a dog crate or large cage, on a clean towel. A heating pad under the towel is beneficial, as is warming the towel in the dryer. Sprigs of lavender or a spritz of lavender essential oil can help her to relax, which is critical. Repeat the soak every hour or so until she lays the egg. If she seems overly stressed at being indoors and away from her flock mates, you can try putting her back into an outdoor pool to try to pass the egg that way. Egg binding is a serious condition if the egg isn't laid within 48 hours or so, so quick treatment is of utmost importance.

Eye Issues/Sticky Eye. Eye issues aren't all that common as long as your ducks have regular access to a deep tub or water or a swimming pool, but sometimes you might have a duck suffering eye irritation – such as a scratched cornea from some dirt, straw or dust in her eye or due to the duck trying to remove debris in her eye. Eye irritation can also be caused by dirty litter, overly muddy conditions or the presence of ammonia in the duck house. Over-mating can cause bubbling eyes, since the sinuses and tear ducts are connected and often irritated by the drake holding onto the back of the duck's head while they're mating.

Frequent close inspection (and cuddling) will alert you to health conditions like sticky eye.

If you notice red, bubbling or tearing eyes, first rinse with saline several times a day, change out all the bedding in the duck house, and be sure the duck has access to fresh water deep enough to submerge her whole head. The condition should clear up in several days.

Sticky eye is a condition from which a sticky discharge seeps from the eyes. Most often it manifests itself if insufficient water is available to bathe in. Providing a clean pool will usually clear it up. If, after several days of the saline rinse and the availability of a deep tub of clean water, the eye is still irritated, it could be an eye infection. Treating with goldenseal or chamomile is recommended. A cooled compress of the steeped herbs, or dripping brewed "tea" into the eyes can help.

Feather Pulling. Feather pulling shouldn't be an issue in a happy flock with plenty of space. However, ducklings will often pick at the blood-filled shafts as they grow in their first feathers. Be sure your ducklings aren't crowded and consider replacing your white heat lamp

There are several possible causes for feather pulling.

with a red bulb; it helps reduce pulling. Offer them lots of greens for added protein, including alfalfa, basil, broccoli florets, dill, fennel, mung bean sprouts, oregano, parsley, spinach and tarragon – all excellent protein sources.

Generally, feather pulling in adult ducks is caused by boredom, lack of protein, overcrowding, over-mating, or parasites. If you notice missing feathers or blood on your ducks, consider what might be the root cause. Whether you need to enlarge your run, add protein to your ducks' diet, separate the drake or add more females, or treat for parasites, feather pecking issues are usually easy to resolve. Feathers missing from the back of the ducks' heads or broken wing or back feathers are usually attributable to an over-zealous drake. Feathers missing from the vent area could signal mites or lice (not very common in ducks), which can usually be treated by provided a swimming pool daily for your ducks or dusting the area with food-grade diatomaceous earth.

Frostbite. Ducks are generally not as susceptible to frostbite as other animals because they have a very efficient circulatory system in their legs and feet and don't have combs or wattles like chickens do that often get frostbitten – but they can suffer frostbitten feet if they aren't able to get up off the cold, frozen ground. Putting down straw in the winter in their pen helps prevent frostbite, as does keeping them inside their house when it is abnormally cold. Your duck house needs to be well-ventilated, since its more often moisture that contributes to frostbite than the cold itself. Duck houses shouldn't be heated for the same reason: the heat creates moisture in the house. Not to mention the fire hazard.

If you do suspect frostbite (if a duck is sitting and not moving much, has ice on her feathers or her feet start turning black), bring her inside and let her warm up. Frostbitten feet often have to be amputated, so care needs to be taken to prevent it if at all possible.

If extreme cold is predicted, smear some coconut oil or my frostbite salve on your ducks' feet and legs to help prevent it from occurring or to treat minor cases.

Ducks' excellent circulatory system allows them to enjoy the winter cold. But you need to be watchful for frostbite.

Homemade Frostbite Ointment

What you'll need:

2 ounces beeswax, grated

¾ cup coconut oil

¼ teaspoon liquid vitamin E (helps repair damaged skin)

10 drops calendula essential oil (anti-inflamatory, aids in healing wounds)

10 drops lavender essential oil (relaxant, pain reliever, antibacterial, anti-fungal)

10 drops lemon essential oil (antibacterial, antiviral)

What to do:

Add grated beeswax and coconut oil to a glass bowl set over a pan of boiling water, then reduce heat to low. Once completely melted, remove from heat and stir in the Vitamin E and essential oils until well mixed. Pour into a covered container and cool. Store in a cool, dry place and use as needed.

Gapeworm. While gapeworm is contracted from eating infected earthworms, slugs or snails, or from eating grass containing the worm eggs, healthy adult ducks generally aren't at risk – but ducklings under the age of three months old can be. Coughing or stretching their necks can be an indicator your duck is infected. The worms attach themselves to the trachea or travel into the lungs, causing respiratory distress, difficulty breathing and ultimately death in severe cases. You can help control gapeworm by adding apple cider vinegar to their water (have you figured out by now that it's a good idea to add it for many reasons!), food-grade diatomaceous earth to their feed or offering fresh minced garlic, free-choice, to your ducks. I wouldn't let the fear of your ducks' contracting it keep you from allowing them to free-range and gobble up all the worms and slugs they can find.

Impacted Crop. Although ducks don't technically have a crop like chickens do, impaction can occur in their gizzard or gullet, where their food is stored. This occurs when not enough grit is provided to your ducks to help them digest their food, or worse, if they swallow a long piece of string, baling twine, rubber band or plastic, or even large amounts of long pieces of cut grass. Never feed your ducks cut grass unless it's cut into pieces 2 inches or shorter. If your ducks don't have frequent access to free-range, put out a dish of commercial grit for them to eat free-choice. If you do suspect an impacted crop, i.e., swollen and hard to the touch, gently massage the mass to help break it up; this can help, as can offering grit, a bit of olive oil and plenty of water.

Allowing your ducks to free range enables them to find grit on their own; otherwise you will need to provide it to them.

Parasites (external). External parasites include mites, lice, ticks, fleas or maggots. Generally, healthy ducks who swim regularly will not be affected by any of these external parasites since the water will drown them and with regular preening your ducks will remove any they find. However, be aware that a sick or injured duck who is not allowed to swim or a duck who can't properly preen herself might become infested. It's still prudent to check your ducks periodically for any parasites, especially if you notice any of your ducks scratching excessively or their feathers starting to look ragged. If you do discover an infestation, spray your duck with garlic juice or neem oil, then dust the vent and under the wings where parasites generally congregate with diatomaceous earth. Repeat the treatment every few days for several weeks. Also treat the duck house – discard all bedding, scrub the house down with vinegar and water then dust the floor with diatomaceous earth before adding new straw.

Parasites (internal). Internal worms are not generally a problem for ducks. Furthermore, healthy ducks are able to handle a low worm load and not be affected by it. Worms only become a problem if a duck is not in otherwise good health and strong enough to keep the worm population in check. Signs of worms, in addition to a fecal test: underweight ducks or perpetually soiled or bloody bottoms. Other signs your duck may have worms include lameness, coughing or shaking of the head.

There are several commercial wormers on the market, including Strongid, Flubenvet and Ivermectin – none of which is licensed for use on ducks. I do not recommend pre-emptive worming of ducks nor do I advocate using commercial wormers on your own. If, and only if, you have a vet-confirmed case of worms would I even consider worming and then only under a vet's care.

Instead, keeping your ducks healthy and adding some natural preventives thought to combat worms to your ducks' regular diet is a far more natural and beneficial way to go. Natural wormers include catnip, chamomile, garlic, nasturtium and ground pumpkin seeds.

Ducks will seek out plants that are natural wormers, like chamomile.

Peritonitis/Soft-Shelled Eggs. A diet lacking in calcium can lead to soft-shelled eggs, which means that the duck has inadequate calcium to form the hard shell around the egg yolk and white. It's very important to feed your ducks layer feed and also supplement that with free-choice oyster shell or crushed eggshell. Ducks will leach calcium from their bones to make the eggshells if they don't have enough in their diet, which can lead to easily broken bones and poor duck health.

Ducks need adequate dietary calcium and sun exposure (Vitamin D) to ensure hard eggshells.

In addition to the calcium, Vitamin D is required for maximum calcium absorption, so allowing your ducks time outdoors year-round is very important. Without enough calcium in their diet, ducks might also lack the ability to contract their muscles to lay their eggs. That can lead to egg binding. (See page 94.)

Even more worrisome is egg peritonitis. This occurs when the abdomen becomes inflamed, possibly due to a buildup of egg yolks. If the oviduct rips, soft-shelled eggs or yolks (or sometimes even hard-shelled eggs) can pass into the abdomen, causing an immovable mass there. While a vet can sometimes successfully operate, often this condition proves fatal.

Prolapsed Vent/Oviduct. A prolapse occurs when a portion of the oviduct is pushed outside the body and doesn't retract on its own. This often happens in conjunction with the passing of a soft-shelled egg or in an egg bound situation. More common in young layers, overweight ducks, or ducks who strain to lay extremely large (double yolk) eggs, it can also be genetic and is often recurring. In minor cases, it can be sufficient to gently push the protrusion back in place, then apply some Green Goo, or coconut oil, and sugar for a few days to help tighten the skin tissue and keep it softened while it heals.

Additional treatment should include withholding layer feed and offering instead lots of leafy greens and some milk – and keeping the duck separated in a quiet dark place to try to prevent her from laying eggs for a few days (ensuring the prolapse doesn't recur and to give her time to recover). Certain ducks can be prone to prolapse and might eventually succumb to the condition despite your best efforts, so it's extremely important to keep your ducks a healthy weight and feeding them a balanced diet.

Here are some preventive measures:

- Allow your flock plenty of space for exercise to help them maintain good muscle tone and a healthy weight.

- Limit high calorie treats (such as corn) to prevent obesity.

For duck health, there's no substitute for a balanced diet and plenty of space for exercise.

- Provide free-choice calcium for hard eggshells.

- Give your hens a natural break from laying through the winter for their bodies to rest. By forcing them to lay year-round (by adding artificial light to your coop in the winter), you are putting an unnatural strain on their bodies and reproductive systems.

Prolapsed Penis. Similar to a prolapsed vent, a prolapse in drakes occurs when the corkscrew-shaped penis doesn't retract back into the body after mating. In minor cases, it will eventually retract on its own, or you can gently push it back inside. It can help to treat it as above with Green Goo and sugar to tighten the skin and keep it from getting infected – as well as separating the drake away from the females for awhile. It's very important to provide your drake a daily swim to help keep the area clean and moistened.

If the penis doesn't retract after a time, a vet can often remove it with no adverse effect on the duck, except he won't be able to fertilize eggs any longer. If the penis is only partially prolapsed, the exposed section will eventually die, turn black and fall off. However, the drake should eventually still be able to breed in this condition.

Respiratory Issues. Ducks are not particularly susceptible to respiratory distress, but if adequate water is not provided during feeding times, coughing, squeaking or difficulty breathing can occur. Other times, dust, debris or ammonia build-up in the litter in the duck house can affect breathing. Plenty of water and new bedding can alleviate the distress.

Basil, bee balm, cinnamon, clover, dill, echinacea, rosemary, thyme, and yarrow all aid in respiratory relief. Feeding them fresh free-choice, or dried mixed into the duck feed, can help keep respiratory systems strong. Any duck with a respiratory issue that doesn't show clear signs of improvement or resolve itself after several days or gets progressively worse should be seen by a qualified vet to rule out more serious illness.

Sinus Infections. The male bites and holds onto the back of the duck's head when he mates with her and that's where the duck's sinuses are located, so vigorous mating can result in sinus problems. Puffed cheeks or mucus draining from the nostrils or eyes can signal a sinus issue. Alleviate the infection by treating the duck with VetRx (holding a Q-tip or cotton ball soaked in the liquid up to the duck's nostrils) and providing clean water. Keeping your drake apart from your duck until she recovers is also recommended.

Spraddle Leg. Spraddle Leg is an affliction that leaves ducklings unable to walk and their legs splayed out to the sides. Spraddle leg is generally treatable by hobbling the legs in place with a Band-aid or piece of Vetrap, but is far easier to prevent than treat by avoiding slippery brooder surfaces such as cardboard or newspaper, and instead using textured rubber shelf liner or shavings.

Wet Feather. Wet feather is a condition most common in ducks not allowed regular access to a pool. A duck has a preen gland at base of the tail which produces oil. Each time a duck is in water, she will preen her feathers both during the swim and after, stimulating the oil gland and distributing the oil onto her feathers. If the gland stops producing oil, or if the duck doesn't have regular access to water allowing her to clean off the dirt and mud from her feathers and maintain the oily coating – then her normally-waterproofed feathers will get waterlogged and stay wet the next time the duck does swim or is out in the rain, often leading to chilling. This often will cause the

It is imperative that ducks have access to water to swim in to prevent wet feather condition.

duck to avoid water altogether, which just makes the condition worse.

(Pekins are most susceptible to wet feather, so if you raise Pekins, it's very important to keep their pen as dry and mud-free as possible and be sure they swim often.)

In addition to inadequate water for bathing, causes of wet feather can include overall poor health, a poor diet that doesn't include enough vitamins, lice or other parasites, or a non-working or clogged preen gland.

Treatment should entail bringing the affected duck indoors and washing her in water and Dawn dish detergent, rinsing her well and then blow drying her on a warm setting before putting her back outside. This removes the old oil as well as

Ducks waterproof their feathers by preening, which stimulates the oil glands and distributes the oil onto the feathers.

any dirt or mud from the feathers and gives her a chance to start over. She should be kept from swimming for a few days, but allowed access to a small tub of water; that way, she can begin to splash a bit of water on her head and get back to the preening process. After a few days, short swims can be allowed again until you see that her waterproofing is back.

Depending on the severity of the wet feather, if that doesn't work, often you will need to wait until the duck molts and grows in new feathers. If you do have a duck suffering from wet feather, you will need to limit her pool time, especially in the colder months, and be sure to dry her off whenever she gets wet.

Wry Neck. Wry neck, sometimes also called "stargazing," "crook neck" or "head tilting," is a condition that causes ducklings to be unable to hold their heads up on their own. The affliction can progress to the point that the little one walks backwards or tumbles over on its back, unable to walk at all. It can be fatal if not treated, as the chick or duckling risks becoming dehydrated and malnourished quickly if unable to eat or drink by itself.

Head tilting can sometimes be an early sign of wry neck.

The exact cause of the condition can vary, from genetics to a head injury to a thiamine (Vitamin B1) or Vitamin E deficiency. Ingesting toxins or contracting botulism can also bring about wry neck. Ducklings are especially prone to being affected by toxins such as lead or other metals in their environment, so care should be taken to remove any potentially dangerous substances. Wry neck can also be a symptom of Marek's disease or aspergillosis. Regardless of the cause, you will want to separate the afflicted duckling to be sure it's not getting trampled and is able to eat and drink unimpeded. Hand feeding might be necessary if the duckling is not able to eat and drink on its own. (Dipping the bill into a small dish of water is far safer than using an eyedropper, which can lead to aspiration.)

I have never had a case of wry neck here on our farm despite hatching and raising several batches of ducklings over the years. I have a strong feeling that our success with not having to deal with it is in part due to the variety of herbs and weeds I offer to our flock on a regular basis. They are packed with vitamins and ensure a well-balanced diet for our ducks. I also only buy hatching eggs from reputable sources, to minimize the chances of hatching embryos that have vitamin deficiencies.

Adding B1 and E vitamins to the ducklings' diet can often successfully cure them. If you see evidence of the condition, regardless of the cause, you should immediately step up your ducklings' diet, since a well-rounded diet is the key to good health. Treatment can take weeks or longer; immediate results should not be expected, but the sooner you begin treating, the better chances for success.

Incorporating some molasses into the afflicted duckling's diet is beneficial, as molasses is packed with a variety of vitamins and nutrients. You can help build up your ducklings' B1 supply by adding some brewer's yeast, bran, sunflower seeds or wheat germ to their diet.

Selenium helps boost the effectiveness of Vitamin E, so simply treating with a Vitamin E supplement may not be enough. What works better is adding natural sources of Vitamin E to your flock's diet, as many of them are also a good source of selenium. The following herbs and spices are high in Vitamin E: basil, cayenne powder, cloves, oregano, parsley, sage and thyme. These herbs and spices contain Vitamin E in lesser amounts: caraway, cinnamon, cumin, marjoram and turmeric.

Other sources of Vitamin E include alfalfa, dandelion, nettle, raspberry leaf and rose hips – as well as spinach, sunflower seeds, pumpkins, squash, fish and olive oil.

The best way to feed the spices to your duckling is to sprinkle them into your duckling's feed, or mix some into scrambled eggs or warm oatmeal, or chop the fresh herbs and feed free-choice. Or make an herbal tea by steeping the fresh or dried herbs in boiling water and then cooling it.

If you're planning on hatching eggs from your own ducks, be sure your flock is eating a healthy diet of good-quality layer feed in advance of collecting the eggs you plan on incubating – and incorporate nutritious weeds and herbs into their diet as well.

Some things not to worry about:

Molting/Feather Loss. If you notice your ducks starting to lose their feathers (normally in the fall), you are probably witnessing the normal, yearly process of molting. Molting is the natural replacement of old, dirty or broken feathers with nice new feathers for winter to help keep them warm and dry. The molt generally takes a few weeks and often won't even be noticeable, except for an abundance of discarded feath-

To hasten regrowth of feathers after molting or other feather loss, increase the amount of protein in the diet.

ers on the ground in the pen. The new feather shafts literally push the old feathers out, so you're not likely to see any bare or bald patches or missing feathers. If you are seeing areas of bare skin on your ducks, it's more likely parasites, feather pulling by the other ducks or rough mating.

Regardless of the cause of the feather loss, increasing the protein in their diet is essential for helping your ducks grow in new feathers. Mealworms, crickets or cooked meat scraps are all good sources of protein.

Spotted or Green Bills. You might notice your ducks' bills becoming covered with green spots that almost look like mold. This is actually nothing to worry about; what is actually happening is that the orange coloring that normally covers a duck's normally green bill is fading, allowing the green to show through. The orange color of ducks' bills, legs and feet comes from xyanthophyll, a carotenoid found in alfalfa, basil, carrots, corn, marigolds, pumpkins and other foods. The more of these foods a duck eats, the more orange their bills and feet will turn. Xyanthophyll is also what colors egg yolks.

One trick to figure out who your good layers are is to look at their bills and feet. The bills and feet of your better layers will be paler than your poorer layers and drakes, since much of the carotene they eat has been used in the production of a nice orange egg yolk. If the appearance of the mottled bill bothers you, try adding more xyanthophyll-rich foods to your ducks' diet. Not only will their bills return to a bright orange, their egg yolks will be a more vibrant orange as well and both your ducks and their eggs will be healthier.

Do your ducks' bills look like this? It may look bad, but it's nothing to worry about.

SOURCES OF ESSENTIAL NUTRIENTS

Protein – Red meat and fish, worms, crickets or other bugs

Calcium – Leafy greens, yogurt and parsley

Vitamin D – Be sure your duck spends at least part of the day outside even if she's confined to a cage in the house to heal.

Vitamin K – Kale, broccoli and spinach

Zinc – Spinach, pumpkin seeds and red meat

This is not meant to be a comprehensive medical guide to duck health, but merely to alert you to potential health issues that can occur in a small, backyard flock and how to treat them at home. If your duck is not responding to your at-home treatment, if the symptoms worsen, or if an injury is life-threatening, a trip the vet is in order.

For more complete information or specific treatment needed, consult your vet or Storey's Guide to Raising Ducks, by David Holderread, 2011.

Multiple unexplained deaths within a short period of time should always be a cause for concern and immediately reported to your state avian lab.

Duck First-Aid Kit

It's always prudent to keep a well-stocked first aid kit at the ready. I consider the items listed below essential to any duck first-aid kit. You probably already have most of them in your pantry or kitchen cabinet. They are all natural and they have no side effects. Keep your kit somewhere handy, so if you need to grab it in a hurry you can. With it, you should be equipped to treat nearly anything.

Cornstarch (stops bleeding)

Cornstarch is an effective way to stop bleeding fast. Applied directly to a shallow cut or feather quill that is bleeding, the cornstarch will staunch the blood flow. Regular flour, a cold wet tea bag or the spice turmeric applied to the site will also work. Turmeric is also an antibacterial, so will help prevent the wound from getting infected.

Honey (heals wounds, antiseptic)

Honey has long been used to treat and heal wounds. Its antimicrobial and antibacterial properties make it my to-go wound dressing. It helps speed healing, prevents infection, and reduces swelling and inflammation. Honey can be slathered on any type of cut or skin laceration.

Oregano/Oregano Oil (natural antibiotic)

Oregano oil has been studied as a natural antibiotic for poultry. If you do have an ailing duck who is suffering an infection, such as staph due to a case of bumblefoot, adding a few drops of oregano oil to the water can help take the place of a conventional antibiotic, especially when paired with cinnamon.

Coconut Oil (lubricant)

Coconut oil is helpful if you have an egg-bound duck to lubricate around her vent. It can also be used to slather on dry bills or feet to soften them.

Cinnamon (respiratory relief)

Although ducks don't suffer respiratory issues nearly as often as chickens do, often over-mating can irritate the sinuses and eyes. Ground cinnamon has been studied as a natural treatment for respiratory distress, especially when paired with oregano. Sprinkle a bit into your ducks' feed if you notice any signs of respiratory illness.

Blackstrap Molasses (energy boost, replaces nutrients lost due to bleeding or heat stress; a toxin flush)

Molasses is packed with nutrients, minerals and antioxidants and helps boost immune systems, correct vitamin deficiencies, ease heat stress, and improve the appetite of sick ducks, as well as increase the body's response to other treatments. High in iron, molasses is extremely beneficial to an injured duck who has lost blood, and in fact can be offered to any duck experiencing any type of stress from extreme heat or cold, predator attack, or a rough molt. It's also excellent to give to new ducklings to ease the strain of shipping or hatching and provide them an energy boost – much like sugar water does – but be aware that too much can cause diarrhea. Just a drop or two orally is sufficient.

Due to this flushing action and its antioxidant properties, molasses can also be used as a laxative detox in the case of suspected accidental poisoning or botulism. It can be given orally using an eyedropper, mixed into water or drizzled onto feed or oatmeal. Be sure to offer lots of fresh, clean water any time you are administering molasses. *(Note: any type of molasses will do in a pinch, but blackstrap is the most nutritious.)* No official dosage has been established for ducks, but adding ½ to 4 ounces of molasses per gallon of water is recommended, depending on the severity of the treatment required and cutting back if you notice excessive diarrhea.

Chamomile Tea Bags (soothing to irritated eyes)

Chamomile has anti-inflammatory properties and has long been used to soothe irritated or infected eyes. Simply wet a chamomile tea bag (make sure chamomile is the only ingredient) in warm water and then press to your duck's eye for about 10 minutes several times a day. You can also make your own tea bags using dried chamomile flowers and a muslin sachet. Alternatively, you can brew a chamomile tea and then hold a soaked cloth or cotton ball to the duck's eye.

Saline (to flush debris or clean wound)

Saline solution is something I keep on hand because ducks can get dirt or debris in their eyes, and a squirt of saline often is enough to rinse it out. It's also good to keep a clean "water" source like this handy to flush or clean a cut or wound.

Epsom Salts (to ease sore joints and inflammation; to soften skin)

Epsom salts added to warm water is recommended for soaking a sore leg or to cleanse an injury or cut foot or to soften the skin to prepare to extract a splinter or bumblefoot kernel. Absorbed through the skin, the salts help relieve pain and inflammation and also help improve circulation.

Green Goo (antibacterial salve, heals wounds)

Green Goo is distributed by Sierra Sage Herbs and is available from their website (www.sierrasageherbs.com) It's an all-natural herbal antibacterial and antifungal topical salve and can be used in place of Neosporin to help speed up the healing process and prevent infection of abrasions, cuts or scrapes. It can be used on various types of animals including chickens, dogs and horses and even on humans, so I keep a few tins around at all times.

Supplies

Tweezers
Small pair of sharp scissors
Vetrap/First-aid tape

Gauze
Eyedropper
Neoprene "booties" – to keep feet dry and clean while healing

TEN

ALL ABOUT DUCK EGGS

Backyard ducks, besides being incredibly cute and a natural form of slug and grub control for your yard, also lay some pretty tasty eggs. Delicious eggs are certainly one of the many benefits of keeping backyard ducks.

Depending on the breed and age of the duck, you can usually expect approximately 150 to 220 eggs per year (see the breed chart on page 134 for more information). A duck will lay an egg roughly every 24 hours, year-round. Our ducks consistently outlay our chickens, especially in the winter when the chickens' production declines due to the shorter days; our ducks will continue to lay without any additional supplemental light in their house.

Size and nutritional value. Duck eggs are roughly 30% larger than a medium chicken egg, weighing in at 3 to 3½ ounces. They contain more vitamins than eggs from chickens similarly fed and pastured. Overall, they are more nutritious than chicken eggs. Duck egg

Duck eggs are a delicious benefit from keeping ducks.

yolks do contain more cholesterol than chicken yolks, but in a recent dramatic about-face concerning cholesterol levels, the 2015 Dietary Guidelines Advisory Committee concluded: "Cholesterol is not considered a nutrient of concern for overconsumption." (If you have questions about how many eggs you should be eating you should check with your doctor.)

NUTRITION FACTS

SERVING SIZE (ONE LARGE EGG)	CHICKEN	DUCK
Per Serving	50g	70g
Calories	70	130
Fat	5g	9.6g
Protein	6g	9g
Sodium	70g	102mg
Cholesterol	210mg	618mg
Carbs	0g	1g
% OF RDA		
Vitamin A	4%	9.4%
Calcium	2%	4.5%
Iron	6%	15%
Vitamin C	0%	0%

[Source: USDA SR-21]

Many people who are allergic to chicken eggs find they can eat duck eggs without a problem, and vice-versa. If you have a problem eating chicken eggs, ask your doctor about the possibility of being able to tolerate duck eggs. And if you've been avoiding chicken eggs because they're supposed to be acidic, you might want to investigate some recent claims that duck eggs are an alkaline food; I haven't found solid data to support this.

As for taste, I find duck eggs to be richer and creamier, as well as a bit "eggier." They sometimes do have a somewhat stronger taste than chicken eggs, but since our chickens and ducks eat a similar, if not identical, diet, I find the taste of all our eggs pretty consistent.

Egg collecting and care. Ducks usually lay their eggs in the pre-dawn hours, so by the time you let your ducks out for the day, they will be done laying. This makes it very convenient, both in keeping the eggs from being laid outside, hidden where you can't find them, as well as allowing you to collect a nice fresh basket of eggs for breakfast when you're done with your morning chores.

Since ducks also are prone to burying their eggs under the straw in their nest, this means that the eggs very rarely freeze in the winter. Not only do I collect them shortly after they're laid, they are well insulated in their straw bed.

Due to their thicker shells and membranes, duck eggs stay fresher longer than chicken eggs. As a general rule, eggs shouldn't be washed before being refrigerated, but you may want to anyway; ducks often lay their eggs in the mud or trample their eggs with mucky feet. Soiled eggs can be rinsed under warm water (at least 20 degrees warmer than the eggshell surface) and gently cleaned using a fingernail, stiff brush, rough sponge or old toothbrush to remove any mud or other debris – then refrigerated, where they should last for at least six weeks.

These eggs will be for the family breakfast and maybe some fun baking later on.

You'll get pale green eggs from Runners, Mallards, Magpies and Anconas.

Color. Duck eggs are generally white, but some breeds lay creamy pink or pale green/blue eggs. Different ducks within the same breed can lay white or green eggs; this is different from chicken breeds, since normally all hens within the same breed lay the same color egg. Some ducks – especially Runners, Mallards, Magpies and Anconas – will often lay pale green eggs, while others of the same breed lay white eggs, even if they both hatched from a green egg. The Cayuga duck breed lays a charcoal gray egg. Duck eggs are highly prized by crafters, especially those who do pysanky (Ukranian egg decorating), since the eggs are large and the shells durable and smooth.

Nutrition Comparison:
Backyard Duck Eggs vs. Commercial Duck Eggs

There appears to be a real advantage to backyard duck eggs over commercially produced duck eggs. A 2007 Mother Earth News study found that backyard eggs contain only about 2/3 of the cholesterol of store-bought eggs. They also contain only 3/4 of the saturated fat, up to seven times more beta-carotene, three times more Vitamin E and twice the omega-3s of commercially produced eggs, due to the healthier free-range diet. But the benefits don't have to be limited to flocks that roam free. Be sure to supplement your penned ducks' layer feed with plenty of seeds, grass, greens, insects and worms to reap the same benefits from their eggs.

ELEVEN

COOKING WITH DUCK EGGS

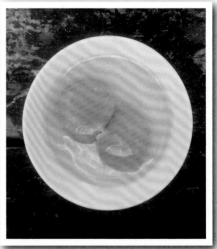

As a rule of thumb, 2 duck eggs are equal to 3 chicken eggs.

Duck eggs are wonderful both for cooking and baking. Being higher in fat content and lower in water content ounce for ounce than chicken eggs – with whites that contain more protein – duck eggs are superior for use in baking. They cause most any baked good to taste richer. Cookies will be more moist and chewy, omelets and quiches will be fluffier and custards creamier. Cakes, cupcakes and breads will rise higher. Bakeries and restaurants that have an extensive dessert menu, Asian markets and high-end restaurants serving locally sourced ingredients, often are interested in buying fresh duck eggs for their kitchens.

For the most part, duck eggs can be prepared the same way you would prepare chicken eggs and substituted into any recipe calling for chicken eggs.

Basic Cooking Tips

Fried or over easy. Duck egg whites tend to get rubbery if they are overcooked, so they should be pan fried very slowly and just to the point of being done.

Poached. Since the whites of duck eggs contain more binding proteins than chicken eggs and fresh egg whites are thicker than the average store-bought egg, you can drop a duck egg right into simmering water to poach it and not worry about any of the typical "tricks" like swirling the water, adding vinegar or salt, using a sieve, cracking the egg into a small cup and then slowly sliding it into the water, etc. I find that our duck eggs poach beautifully.

Hard-boiled. Fresh eggs don't peel well. Air has not had time to penetrate the shell through the pores and separate the two inner membranes to create an air pocket between them. This air pocket is what facilitates peeling, so the older the egg, the easier it is to peel. Instead of letting your eggs "age" before hard cooking them, try steaming them. I use a bamboo steamer over a pot of boiling water, but you can use a metal vegetable steamer basket or even a double boiler. Steam the eggs for twenty minutes, then immediately submerge them in a bowl of ice water until cool enough to peel. Perfectly peeled fresh eggs every time – even those laid that morning.

Scrambled or in an omelet. Scrambled duck eggs are wonderful with some fresh, chopped herbs such as dill, basil or tarragon mixed in with some cheese –and the same goes for making omelets. With their slightly stronger, "'eggy'" taste, duck eggs are wonderful when paired with a strong cheese, like Roquefort or sharp cheddar. Owing to their higher fat content, they will be creamy and tasty even without adding any milk or cream.

Meringues. Duck egg whites don't whip up as well as chicken egg whites, but letting them come to room temperature and adding a bit of lemon juice helps them hold their shape.

Due to their somewhat richer taste, I find that duck eggs pair extremely well with the bold flavor of fresh herbs. I am excited to share with you some of my favorite recipes using fresh duck eggs from the coop – many using fresh herbs from the garden. It's time for fresh duck eggs to shine!

A word of caution: Be aware that consuming partially cooked eggs can pose a health risk to children, the elderly, those with compromised immune systems, and during pregnancy.

Lemon Fried Eggs

(serves 2)

When you have fresh eggs, the simplest of recipes with the fewest ingredients are often the best to let the taste of your eggs shine. This recipe requires fresh eggs, fresh lemon juice and the best-quality olive oil you can find. And maybe a pinch of salt. I like my egg yolks cooked through (not runny), so I break the yolks and flip the eggs over halfway through, but you can do eggs sunnyside up using this method as well, if that is your preference.

What you'll need:

3-4 duck eggs
3-4 tablespoons good-quality olive oil
Slice fresh lemon
Sea salt, optional

What to do:

Heat a cast iron or nonstick skillet over medium-high heat. Once the pan is heated, swirl in enough olive oil to cover the bottom of the pan, plus make a small pool when you tilt the pan (about 1 tablespoon per egg – be generous!) Add the lemon slice.

Give the oil a few seconds to heat up, then carefully crack the eggs into the pan – the oil may splatter – and sprinkle the yolks with a pinch of salt if desired. Cook, tilting the pan and using a spoon to baste the tops of the eggs with olive oil.

When the whites are puffed and set, and the edges are browned and crispy, remove from heat (or flip the egg with a spatula, breaking the yolk if desired, and cook until the other side is done). Squeeze the lemon slice over the eggs, season with additional salt to taste and slide the eggs onto your plate.

Shirred Eggs with Tarragon

(serves 2)

Shirred eggs, or oeufs en cocotte, is an old French way of baking eggs in individual dishes or "cocottes." The richness of the duck eggs is enhanced by the fresh butter and heavy cream in this recipe. Due to the ease of cooking and sophisticated presentation, this is a wonderful way to prepare eggs for a large group for brunch just by doubling or tripling the recipe. There are many variations which use herbs, cheeses, even meats in the ramekins, but I prefer to keep it simple with just a bit of fresh tarragon (an herb that pairs very well with eggs) and some buttered bread crumbs on top.

What you'll need:

4 duck eggs
2 tablespoons heavy cream
Fresh tarragon, coarsely chopped
Salt and white pepper
2 tablespoons panko bread crumbs
½ teaspoon butter, plus more to
   grease ramekins

What to do:

Preheat oven to 325 degrees F.

Lightly butter the inside of the ramekins. Pour the cream around the edges, then break two eggs side by side into each dish. Sprinkle with tarragon and season with salt and pepper. Melt the ½ teaspoon butter and stir in the breadcrumbs with a fork to combine. Top the eggs with the crumbs. Bake for 16-18 minutes or until the whites are just set. For a firmer yolk, bake for several minutes more. Let sit for several minutes, then serve warm.

Eggs Benedict

(serves 2)

Eggs Benedict makes for a truly decadent breakfast, but it really isn't that hard at all. Although mastering Hollandaise sauce can be a bit tricky because it tends to "break" or separate, once you've perfected your technique, the most difficult thing about making eggs Benedict is the timing of it all – so your eggs, English muffins and sauce are all done at the same time and nothing gets overcooked, burned or cold.

Some recipes use Canadian bacon, spinach or other additional ingredients, but I prefer to keep it simple and really highlight our fresh eggs, both poached and in the delicious, rich sauce.

What you'll need:

4 duck eggs, plus 3 yolks for the sauce
2 split English muffins
2 tablespoons fresh lemon juice
1 tablespoon water
1 stick cold salted butter, cut into small pieces
Salt and white pepper
Fresh grated nutmeg

What to do:

Poach the 4 eggs in simmering water until soft set. Keep warm. Toast the English muffins and keep warm.

Meanwhile, for the sauce, whisk 3 egg yolks, the lemon juice and water in a heatproof glass (not metal) bowl set over a pot of boiling water.

Slowly add the butter, a few pieces at a time, until it's all incorporated, whisking continuously. Continue to cook for another minute or two until sauce thickens. Season with salt and white pepper to taste. Immediately remove from heat. Place an egg on each muffin half, cover with sauce, and grate fresh nutmeg on top. Serve any remaining sauce on the side.

Broccoli Cheddar Tart

(serves 6-8)

I tend to make frittatas more often than quiches or tarts, purely because they are quicker and easier, not requiring a crust. But sometimes I just feel like making the extra effort and actually enjoy the relaxation and challenge of making and rolling out a crust. This tart is one of my favorites and works equally well for breakfast, brunch or a light summer dinner with a salad. You can use a traditional round tart pan, rectangle pan or even a regular pie plate.

What you'll need:

1 crust (blind baked)
½ cup broccoli florets,
   chopped into small pieces
¼ cup chopped onion
2 garlic cloves, sliced thin
3 duck eggs

1 cup heavy cream
4 breakfast sausages, sliced
¾ cups shredded
   Cheddar cheese
2 teaspoons flour
Salt, pepper and nutmeg

What to do:

I like to use a pate brisee crust for this tart, but you can substitute your favorite crust recipe or use a store-bought crust. To blind bake, I chill for at least 30 minutes once I've rolled out the dough and have it in the pan. I prick the bottom, then line it with parchment and pie weights and bake the crust at 400 degrees F for 20-30 minutes until it is lightly browned and nearly done. Then remove from the oven and let cool, reducing the oven temperature to 325 degrees.

Meanwhile, cook the sausage until browned and crispy in a cast-iron skillet. Remove and roast the broccoli, onion and garlic in the same pan over high heat until crispy and browned.

Whisk the eggs and heavy cream, season with salt, pepper and nutmeg. In a separate bowl, toss together the shredded cheese and flour.

Arrange the broccoli mixture in the tart pan, cover with the cheese and then carefully pour the egg mixture over the top. Bake for 20-30 minutes until just set. Cover the crust with foil, if necessary, to prevent over-browning. Remove from the oven, cut into slices or wedges and serve warm or at room temperature.

Thyme and Swiss Cheese Souffles

(serves 4)

What you'll need:

1¼ cup milk

¼ cup heavy cream

Sprig of fresh thyme

3 tablespoons butter

¼ cup plus 1 tablespoon flour

4 duck eggs, separated,
  at room temperature

¼ cup freshly grated Parmesan cheese

½ teaspoon salt

Freshly grated nutmeg

⅛ teaspoon fresh lemon juice

¾ cup diced Swiss cheese

4 6-ounce ramekins, cocottes or
other heat-proof stoneware dishes

What to do:

Preheat oven to 350 degrees F.

Lightly coat the inside of each ramekin with butter or cooking spray. Refrigerate until ready to fill.

Bring milk, cream and thyme sprig to a simmer in a small saucepan over medium-low heat, stirring occasionally. Meanwhile, melt butter in a skillet over medium heat, then add the flour, stirring constantly for about a minute until the mixture foams. Whisk in the warm milk and continue to whisk for another minute, until roux begins to thicken. Remove the thyme, pour the mixture into a large bowl and whisk in two of the egg yolks (discard the remaining two yolks or use them for another recipe), the Parmesan cheese, salt and nutmeg. Set aside to cool.

In a large bowl, beat the four egg whites and lemon juice on low until foamy. Increase the speed to high and beat until soft peaks form. Gently fold 1/4 of the egg whites into the roux with a spatula, then carefully fold in the remaining egg whites until just blended. Add the Swiss cheese and gently fold into the mixture.

Divide the mixture among the prepared ramekins and bake for 30 minutes, or until puffed and golden on top. Serve immediately, using additional thyme as garnish.

Sweet Potato Sausage Frittata

(serves 6-8)

I call this recipe my "winter" frittata because it uses some fall crops and it's a bit heavier than most "summer" frittatas (which often feature tomatoes, zucchini or yellow squash). I grow sage, onions and sweet potatoes in my garden and love to use them all in this delicious dish. Equally good for breakfast, brunch, lunch or dinner, frittatas are a wonderful way to use up some of your duck eggs. Omit the meat if you are a vegetarian; substitute zucchini for the sweet potatoes. It's all up to you.

What you'll need:

4 tablespoons olive oil
½ large onion, sliced thin
2 sausages, casings removed and crumbled (I like to use Italian sweet sausage)
1 large sweet potato, cut into thin slices
Handful of fresh sage leaves, julienned, with a few reserved for garnish
12 duck eggs
½ cup heavy cream
1 cup shredded cheese of your choice, divided (I like to use Gruyere or a nice mild cheese like
  Gouda or fontina)
Sea salt
Fresh cracked black pepper

What to do:

Heat 2 tablespoons of the olive oil in a cast-iron skillet over medium-low heat. Add the onions and season with salt and pepper. Cook, stirring occasionally, until golden brown and caramelized, about 30 minutes. Remove onions from pan and add the sausage pieces. Cook, continuing to stir, until the sausage is cooked through and lightly browned. Drain and set aside.

Meanwhile, preheat oven to 400 degrees F.

Toss the sweet potatoes with the remaining 2 tablespoons of olive oil and arrange the slices in a single layer on a baking sheet. Roast for 15-20 minutes, turning the slices halfway through. When the slices are fork-soft and starting to brown, remove the baking sheet from the oven and drain any excess oil.

Reduce the oven heat to 350 degrees. Whisk the cream and eggs, stir in the sage leaves and season with salt and pepper. Place the roasted potatoes, sausage, onions and half the cheese in the skillet, then pour the egg mixture over the top. Bake for 25-30 minutes until the egg is puffed and set. Sprinkle the top with cheese. Let sit, or return to the oven, until the cheese is melted. Slice into wedges and serve warm or at room temperature with a garnish of fresh sage leaves.

Herbed Deviled Eggs

(makes one dozen)

What you'll need:

6 duck eggs, hard-cooked, peeled and halved

⅓ cup mayonnaise (see recipe for homemade
   mayonnaise below)

1 tablespoon Champagne vinaigrette

Pinch of sugar

Salt and white pepper

Fresh chopped herbs – I particularly enjoy a mixture
   of dill, parsley and basil

What to do:

Remove the egg yolks and mash them in a large bowl with a fork or pastry cutter. Add the remaining ingredients and mash until combined. Arrange the whites on a dish and scoop the yolk mixture into the halved egg whites with a small ice cream scoop. Garnish with additional fresh herbs.

Homemade Mayonnaise

(makes about 1¼ cups)

What you'll need:

1 duck egg

Pinch of salt

¼ teaspoon Champagne vinegar

½ lemon, juiced

¾ cup cooking oil

¼ cup olive oil

What to do:

Break the egg into a blender and add the salt. Blend on high 2-3 seconds until foamy. Add the vinegar and lemon juice and blend for several seconds longer. On high speed, with the cover on and the little insert in the center of the lid removed, slowly pour the combined oils into the blender in a slow, steady stream. The mixture should thicken. It will last in the refrigerator for about a week.

Homemade Pasta

(makes about 12 ounces; serves 2-3)

Fresh pasta is incredibly easy and quick to make and the taste is out of this world, especially when you use duck eggs. I love to pair freshly made fettuccine with a nice carbonara sauce and extra Parmesan cheese. I think the rich, creamy sauce is the perfect complement to this fresh pasta.

What you'll need:

2 cups flour (any kind works, I just use regular all-purpose flour)

2 duck eggs, plus one yolk

What to do:

Measure the flour out onto a clean work surface (the counter, kitchen table or a cutting board), making a well in the middle of the mound. Pour the eggs into the mound and then with a fork, work the egg into the flour to create a dough. Once you have the flour incorporated, continue kneading the dough with your hands for about 10 minutes until it is firm and smooth. Form the dough into a ball and then slightly flatten it and wrap it in plastic wrap. Let the dough rest on the counter for at least 30 minutes, and up to several hours, until you are ready to cook it.

Unwrap the dough and roll it out and cut it or use a pasta machine to roll and cut your shapes. Add to a pot of generously salted boiling water and cook for several minutes until al dente, cooked but slightly firm when you bite it. Drain and serve with your favorite sauce. Uncooked pasta can be hung and dried to cook later, or even frozen.

Lemony Egg Rice Soup

(serves 4-6)

This soup is so easy to make and uses basic ingredients that I always have on hand. The zesty lemon really perks up the plain rice and eggs.

What you'll need:

2 quarts chicken broth
½ cup rice (brown or white, both
   work just fine)
4 duck eggs
Zest and juice from one lemon,
   reserve some for garnish
Splash of heavy cream
Salt and pepper
Fresh basil, julienned, for garnish
Drizzle of good quality olive oil

What to do:

Bring rice and broth to a boil in a large pot, then turn heat down and simmer for 15-20 minutes, or until rice is done. Remove from heat, add the lemon zest and season with salt and pepper to taste.

Whisk the eggs and lemon juice in a small bowl and then whisk in a few ladles of the soup, one ladle at a time so the eggs don't curdle. Add the egg mixture to the soup, add a splash of heavy cream, return to the heat and stir over low until just warmed. Garnish with basil and reserved lemon zest. Finish with a drizzle of olive oil, if desired. Serve immediately.

Creme Brulee

(makes 4)

What you'll need:

2 cups heavy cream
⅓ cup sugar
1 vanilla bean, scraped
3 duck egg yolks
Pinch of salt

4 6-ounce ramekins, cocottes or
    other heat-proof stoneware dishes
4 tablespoons vanilla bean-infused sugar
    (make several weeks ahead)

What to do:

Preheat oven to 300 degrees F.

In a heavy saucepan, heat the cream, half of the sugar and the vanilla bean to a simmer over low heat until bubbles just start to form around the edges of the pan.

Meanwhile, whisk the egg yolks and remaining sugar and salt. Set aside.

Remove the cream mixture from the heat and ladle slowly into the egg mixture, whisking so the eggs don't curdle. Once incorporated, strain the liquid into a measuring cup.

Divide the liquid among the ramekins, set them in a baking dish or on a roasting pan. Set the pan in the oven and pour boiling water into the pan halfway up the sides of the ramekins. Bake for 40 minutes. Chill at least 3 hours (or overnight).

Remove from the refrigerator and sprinkle each custard with 1 tablespoon vanilla bean-infused sugar. Brown with a kitchen torch or broil until the sugar melts and bubbles (2-3 minutes). Let cool to harden a bit and serve.

To make vanilla sugar: Split a vanilla bean and scrape out the seeds into a Mason jar. Add the bean, fill the jar with sugar and stir to mix the contents. Let sit for several weeks to allow the vanilla bean to infuse, stirring occasionally.

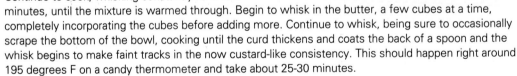

Pink Grapefruit Vanilla Curd
(makes about 3 cups)

Lemon curd is a classic topping usually found atop toast, pancakes, English muffins or scones. This version, which uses pink grapefruit instead of lemons and adds some vanilla bean, is made even more delicious with fresh duck eggs. It can also be used as a donut, macaron or cupcake filling, a tart base, filling for a layer cake or even folded into some whipped cream for a delicious cake frosting.

What you'll need:

6 duck eggs
1½ cups granulated sugar
½ cup fresh-squeezed pink grapefruit juice
   (about ½ grapefruit)
1 vanilla bean, seeds scraped
1 stick (8 tablespoons) unsalted butter at room
   temperature, cut into cubes

What to do:

Set a medium glass bowl over a saucepan of boiling water. Whisk the eggs, sugar and vanilla bean together in the bowl, then whisk in the citrus juice.

Continue to cook, whisking constantly for a few minutes, until the mixture is warmed through. Begin to whisk in the butter, a few cubes at a time, completely incorporating the cubes before adding more. Continue to whisk, being sure to occasionally scrape the bottom of the bowl, cooking until the curd thickens and coats the back of a spoon and the whisk begins to make faint tracks in the now custard-like consistency. This should happen right around 195 degrees F on a candy thermometer and take about 25-30 minutes.

Remove from heat and let cool. Store in a covered container or in the bowl with a piece of plastic wrap over the top. The curd will keep for about two weeks in the refrigerator – although it's so good, and so easy to just eat off a spoon, I doubt it will actually last that long!

Vanilla Bean Cheesecake with Lemon Curd Topping

(makes 8 servings)

Cheesecake made using duck eggs is extremely creamy and rich without needing to add any sour cream. For a citrusy lemon curd topping you can use the Pink Grapefruit Vanilla Curd recipe opposite, substituting ½ cup fresh lemon juice (about 2 lemons) for the grapefruit.

What you'll need:

1½ cups shortbread cookies, crushed into crumbs

6 tablespoons unsalted butter, melted

4 8-ounce packages cream cheese at room temperature

1 cup sugar

3 duck eggs

2 teaspoons vanilla bean paste, or equal amount vanilla extract

Zest from one lemon

1 cup lemon curd

Fresh violets and mint leaves for garnish

What to do:

Preheat oven to 325 degrees F.

In a small bowl, mix the cookie crumbs and melted butter with a fork until combined. Press into bottom and an inch up sides of a 9-inch springform pan. Refrigerate for 5 minutes. Meanwhile, using a stand mixer, beat the cream cheese and sugar until smooth and fluffy. Beat in the eggs, one at a time, and then the vanilla bean paste and lemon zest. Scrape down the sides of the bowl and continue to mix until well blended. Pour the batter into the prepared crust and lightly tap the pan on the counter to remove any bubbles and smooth the top.

Wrap the bottom and sides of the springform pan in aluminum foil, then place in a broiler pan on the bottom rack of the oven. Carefully pour boiling water into the broiler pan so the water comes halfway up the sides of the springform pan.

Bake for one hour, or until edges begin to puff and the center is almost set, but jiggles a bit when shaken. Remove the cheesecake from the oven and allow to cool. Spread the lemon curd over the top and refrigerate the cheesecake for at least 4 hours and up to 2 days.

Just before serving, run a sharp knife around the edge to loosen and release the side of the springform pan. Garnish with mint leaves and violets.

Fresh Mint Chip Ice Cream

(makes about 2 quarts)

What better use for your fresh eggs in the summer than in homemade ice cream? I love this recipe because it not only uses fresh eggs, but also fresh mint leaves – and we have an abundance of both this time of year. The cool, crisp flavor of fresh mint mingles with dark chocolate in this rich, velvety, creamy ice cream. This recipe is a keeper! There is nothing like homemade ice cream on a hot summer evening.

What you'll need:

4 cups milk

2 cups heavy cream

Generous handful fresh mint leaves, ripped
   into pieces

1½ cups sugar

½ teaspoon salt

4 duck egg yolks

2 teaspoons vanilla bean paste or 2 vanilla
   beans, seeds scraped

½ cup chopped dark chocolate

What to do:

Simmer the milk, cream and mint leaves in a saucepan over medium heat just until bubbles form around the edges. Remove from heat and let steep for 15 minutes, then strain through a mesh colander, pressing the liquid out of the mint leaves, and return the liquid to the pan over medium heat.

Meanwhile, whisk the sugar, salt and egg yolks in a medium bowl. Slowly add some of the warm milk mixture to the egg mixture, whisking to combine, then pour the egg mixture into the saucepan with the remaining milk mixture and cook for 2-3 minutes over medium-low heat, whisking constantly.

Remove from heat, whisk in the vanilla, and pour the mixture into a bowl. Refrigerate until completely cooled. Once cooled, stir in the chocolate pieces, pour the mixture into your ice cream maker and follow the manufacturer's instructions.

Appendix

DUCK BREEDS AT A GLANCE

Although all ducks make great additions to your backyard flock – and all lay eggs, eat bugs and weeds and will provide you hours of entertainment – certain breeds to tend to excel in one area or another. You might be drawn to a particular look or temperament, you might want a breed that tends to go broody or is a better forager. Here's a handy reference chart summarizing fourteen of the more popular duck breeds and their traits.

FOR EGGS

KHAKI CAMPBELL
Egg color: white to cream
Lay rate: 250-340 per year
Weight: 4-4.5 lbs.
Temperament: docile, active
Broody: no

SILVER APPLEYARD
Egg color: white
Lay rate: 200-270 per year
Weight: 8-9 lbs.
Temperament: docile, calm, active
Broody: fair/good

WELSH HARLEQUIN
Egg color: white, green, blue
Lay rate: 240-330 per year
Weight: 5-6 lbs.
Temperament: docile, active
Broody: poor

FORAGERS

ANCONA
Egg color: white, cream, blue or green
Lay rate: 210-280 per year
Weight: 6-6.5 lbs.
Temperament: calm, active
Broody: fair/good

CAYUGA
Egg color: grayish/black to white
Lay rate: 100-150 per year
Weight: 7-8 lbs.
Temperament: docile
Broody: yes

MAGPIE
Egg color: white
Lay rate: 220-290 per year
Weight: 5.5-6 lbs.
Temperament: quiet, docile, active
Broody: fair/good

FOR BROODING

DUTCH HOOKBILL
Egg color: white, blue or green
Lay rate: 100-225 per year
Weight: 3.5-4 lbs.
Temperament: calm, docile, active
Broody: yes

MALLARD
Egg color: green/blue
Lay rate: 60-120 per year
Weight: 2-2.5 lbs.
Temperament: calm
Broody: yes

ROUEN
Egg color: white
Lay rate: 30-125 lbs.
Weight: 8-10 lbs.
Temperament: docile, calm, good foragers
Broody: yes

FOR PEST CONTROL

RUNNERS
Egg color: white, blue, green
Lay rate: 200+ per year
Weight: 4-4.5 lbs.
Temperament: docile, can be excitable, originally used in Asia in rice paddies to control insects, frogs, snakes and lizards.
Broody: no

ALL-PURPOSE PETS

BUFF ORPINGTON
Egg color: white, cream or blue
Lay rate: 150-220 per year
Weight: 7-8 lbs.
Temperament: calm, active
Broody: fair/good

PEKIN
Egg color: white
Lay rate: 125-225 per year
Weight: 8-10 lbs.
Temperament: friendly, calm, not great foragers
Broody: no

SAXONY
Egg color: white, blue or green
Lay rate: 190-240 per year
Weight: 8-9 lbs.
Temperament: docile
Broody: fair/good

SWEDISH
Egg color: white, cream, green
Lay rate: 100-150 per year
Weight: 6.5-8 lbs.
Temperament: docile, active, hardy, good forager
Broody: fair/good

EDIBLE HERBS, WEEDS AND FLOWERS

I have long believed that ducks, like other animals, instinctively know what they need from nature and if allowed to roam freely or offered a variety of herbs and weeds free-choice, they will eat as much or as little as necessary for optimal health. My hypothesis was tested last summer and I do honestly believe it proved my theory.

Our Pekin drake, Gregory, like other drakes (and roosters too), will stand back when treats are offered and let the females eat first. Whether it's good old-fashioned chivalry or the fact that the males don't have as high nutritional requirements since they aren't expending energy laying eggs, I have watched time and time again as Gregory lets the other ducks eat their fill and then takes his turn.

On this one occasion, Gregory was limping, most likely having hurt his foot chasing the girls around the run (it was mating season after all!), and when I returned to the pen having picked a large pail of chickweed, Gregory, behaving as I have never witnessed before, literally shoved the ducks aside and started gobbling up the chickweed as soon as I put it on the ground. I was confused by his behavior only for a moment, until I remember that chickweed, among other things, is a natural pain reliever and anti-inflammatory. I realized that Gregory knew he needed it right then more than the ducks did. I continued to pick some chickweed daily for several days knowing it would help Gregory feel better and his leg heal, and he continued to feast on it. In several days, he was walking normally again and he went back to his usual chivalrous behavior of "ladies first." Since that time, I have never again seen him push to eat first. So...do animals possess the wherewithal to heal themselves if provided the right tools? I sure am a believer.

Ducks love to eat anything green. That includes all kinds of weeds, herbs, edible flowers, grasses and leafy greens from the garden. Super nutritious and economical, these treats make up a large part of what I supplement my ducks' daily diet of layer feed with – especially beneficial if your ducks are penned up and don't free-range full time.

I make it a point to pick weeds nearly every day for our ducks, and each spring I plant them their very own herb garden. They particularly prefer water-based herbs over the woody herbs (think parsley and basil instead of rosemary and thyme), but I do grow all kinds of herbs for our ducks. I add them dried into their feed year-round. Starting your ducklings off young trying different foods will help prevent picky eaters as they grow into adult ducks, so offer fresh weeds and herbs to your young ducks early on. Ducks also love to eat flowers! You'll need to keep them out of your landscaping and gardens or they'll decimate all your pretty blooms, but do consider planting some edible flowers specifically for your ducks.

Just take a look at some of the health benefits of all these favorite duck treats. It's no wonder that ducks love to eat them! Some common plants that are my ducks' favorites include:

WEEDS

Bedstraw – aids in mucus membrane health, antioxidant, calming, aids in thyroid function, effective blood clotting agent

Bittercress (Shotweed) – high in Vitamins C and A and beta carotene, fights infection

Chickweed – high in Vitamins A, B and C, anti-inflammatory, pain reliever, supports mucus membrane and digestive health, source of omega-6

Clover (leaves and flowers) – high in Vitamins A and B, calcium, iron and protein, detoxifier, respiratory health, appetite stimulant

Dandelion (leaves and flowers) – high in Vitamins A, C and K, iron and calcium, strengthens the immune system, antioxidant, detoxifier, health tonic

Plantain – cough suppressant, aids respiratory health, supports healthy mucus membranes, thought to be a natural wormer

Purple Dead Nettle (Henbit) – high in fiber and iron, blood detoxifier, eases muscle pain, anti-inflammatory, antimicrobial, fights staph and other infections, eases respiratory infections, aids in digestive health

Purslane – high in fiber, vitamins A, B and C and omega-3s, aids in respiratory health, good source of niacin, antioxidant, aids in digestive health

Red Raspberry (leaves and fruit) – high in Vitamins A, B, K, niacin and fiber, boosts fertility and reproductive system health, aids in cardiac health, antioxidant, anti-inflammatory, fights infection

Smartweed (Heart's Ease/Lady's Thumb) – antioxidant, aids in respiratory health, antibacterial, anti-inflammatory

Wild Strawberry (leaves and fruit) – detoxifier, high in iron, overall health tonic

HERBS

Basil – antibacterial, supports mucus membrane health, fights infection, calming, general tonic

Catnip/Catmint – high in Vitamin C, calming, aids in digestive health

Chervil – high in Vitamin C, carotene for orange egg yolks, feet and bills, digestive and circulatory system aid

Cilantro – high in Vitamins A (vision) and K (clotting), antioxidant, assists in bone development

Dill – supports digestive and respiratory health, antioxidant, relaxant

Lemon Balm – calming, antibacterial, aids in digestion

Marjoram – laying stimulant, calming, digestive aid

Mint – anti-oxidant, aids respiratory and digestive health, naturally lowers body temperatures

Nettle – high in calcium, protein and other nutrients, health tonic, aids in digestive health,

Oregano – supports the immune system, being studied as a natural antibiotic

Parsley – high in many vitamins, specifically Vitamin B, aids in blood vessel development, laying stimulant, digestive aid, antiseptic, supports respiratory health

Sage – antioxidant, antiparasitic, supports overall health, calming, thought to help battle salmonella, aids in digestive health, antiseptic, aids in mucus membrane and respiratory health

Tarragon – high in vitamins A and C, antioxidant, appetite stimulant, aids overall health, aids in digestion

EDIBLE FLOWERS

Bee Balm/Bergamot/Monarda (leaves and petals) – antiseptic, supports respiratory health, calming, digestive aid, stimulant

Chamomile (flowers) – general tonic, heals wounds

Echinacea (leaves and petals) – supports respiratory health, strengthens immune system, antibacterial, antiviral

Marigold/Calendula (flowers) – makes vibrant orange egg yolks, bills and feet, detoxifier, aids digestion, anti-inflammatory, aids in healing of skin and wounds

Nasturtium (flowers and leaves) – high in Vitamin C, laying stimulant, antiseptic, supports respiratory health, natural wormer

Roses (petals and hips) – high in Vitamin C, antiseptic, antibacterial, detoxifier, aids circulatory system, calming, steeped petals can help soothe eye infections

Squash blossoms – high in calcium, iron and Vitamin A

Sunflowers (petals) – anti-inflammatory, cough suppressant (seeds are high in vitamin B, niacin, iron and protein)

Violets (petals and leaves) – high in Vitamin C, aid circulatory system, anti-inflammatory

Yarrow (flowers and leaves) – antibacterial, anti-inflammatory supports digestive and respiratory systems and mucus membrane health

ACKNOWLEDGMENTS

I would like to thank my Mom – not only for choosing the country life for our family and making me the country girl that I am, but also for always believing that I can do anything I set out to do. I am a far better person because of your constant support and confidence in me.

Thanks are in order to my husband, Mark, for deciding that yes, we did need a few ducklings in addition to the baby chicks we were picking up at the feed store; and for being supportive of my endeavors over the years and only sometimes rolling his eyes when I tell him that I am hatching more chicks or ducklings.

I also need to thank our two dogs, Bella and Winston, for being so patient with me while I was writing this book, even though I know they wanted me to play with them. And more importantly, for never, ever chasing the ducks – but instead acting as their guardians, even chasing off foxes and rabid raccoons.

Special thanks to everyone who contributed photos to this book, including Erica Deeds Kellam, Sarah Barrera, Tracy at Duck Duck Dog, Brian Baisch, David Allen Cooper, Ann at A Farm Girl in the Making, Chloe's Creek, Amy Fewell, Tonya Mappin, Leigh Shilling Edwards

and Darlene Terry. Your talents helped make this book even better than I could have done on my own.

Thank you to all those who follow the ducks' adventures on Facebook and the blog. This book became a reality in large part due to you.

And finally, once again, a round of applause for the team at St. Lynn's Press – Paul, Cathy, Holly and Chloe. Yet again, you managed to take my rough draft, thoughts and photos and turn them into a book that is far greater than the sum of its parts.

INDEX

INDEX

REFERENCES/FURTHER READING

Choosing and Keeping Ducks, by Liz Wright,
   TFH Publications Inc, 2008

*Ducks: Tending a Small-Scale Flock for Pleasure
   and Profit,* by Cherie Langlois, Hobby Farms
   Press, 2008

Herbal Medicine, by Dian Dincin Buchman,
   David McKay Company, Inc., 1979

Storey's Guide to Raising Ducks, by Dave
   Holderread, Storey Publishing, 2011

*The Complete Book of Herbs: A Practical Guide to
   Growing and Using Herbs,* by Lesley Bremness,
   Viking Penguin, 1988

The Domestic Duck, by Chris and Mike Ashton,
   The Crowood Press Ltd., 2001

The Herb Book, by John Lust, Dover Publications,
   Inc., 1974

The Ultimate Pet Duck Guidebook,
   by Kimberly Link, 2014

Online Resources

For health and other issues relating to backyard
ducks, these two websites are my preferred
resources:

- Cornell University College of Veterinary Medicine,
  Duck Research Laboratory, www.duckhealth.com

- Metzer Farms, www.metzerfarms.com

Rare Breed Ducks

Many of the breeds of ducks featured in this book
are currently on the endangered or critical lists.
Consider choosing a rare breed of duck. These two
wonderful organizations provide lots of information
on the various breeds and their status, as well as
help in finding a breeder.

- The Livestock Conservancy,
  www.livestockconservancy.org

- Heritage Poultry Conservancy, founded by
  P. Allen Smith, www.heritagepoultry.org

Rescue a Duck

Sadly, many ducks are abandoned after Easter or
when they are no longer wanted. Domestic ducks
can't fly to escape predators or to migrate south
to escape the cold and find food in the winter, and
many end up starving or being harmed by wild
animals, dogs or even humans. Consider rescuing
an abandoned duck (or two or three!) from one of
these duck rescue organizations:

- Carolina Waterfowl Rescue, Indian Trail, NC;
  www.carolinawaterfowlrescue.com

- Chloe's Creek, Mount Joy, PA; Chloe's Creek on
  Facebook.com

- Majestic Waterfowl Sanctuary, Lebanon, CT;
  www.majesticwaterfowl.org

ABOUT THE AUTHOR

LISA STEELE is one of the most trusted voices in small-flock poultry keeping. Her first book, *Fresh Eggs Daily*, was all about healthy, natural care for chickens – a unique approach to chicken keeping, using a blend of old-timers wisdom and Lisa's original tried-and-true methods. Now comes *Duck Eggs Daily*, a companion guide for the backyard duck enthusiast. Lisa's own happy, healthy ducks benefit from her wide knowledge of herbs and other natural health enhancers and preventives – which she is delighted to share.

Since 2009, when she brought home her first two fluffy ducklings, Puddles and Bob, she has focused on caring for them, and those that have followed, as naturally as possible. Her methods have attracted a loyal readership on her popular award-winning website Fresh Eggs Daily (www.fresheggsdaily.com), named one of the Top Ten Garden Blogs of 2014 by *Better Homes and Gardens* magazine. Lisa is a frequent contributor to *Backyard Poultry*, *Hobby Farms* and *Chickens* magazines. She has been featured on HGTV

Gardens, DIY Network and Country Woman online and frequently appears on various television and radio shows nationwide talking about raising chickens and ducks naturally. Most recently, she is hosting her own TV show called Welcome to my Farm, airing on NBC in Maine and streaming online.

Lisa lives on a small hobby farm in Maine with her husband and ever-growing flock of ducks, chickens and assorted four-legged friends. In her free time she enjoys cooking with produce fresh from her garden and baking with duck eggs fresh from the backyard. She invites you to join her on Facebook and Instagram where she shares farm photos and advice under the name Fresh Eggs Daily.

Other books from St. Lynn's Press

www.stlynnspress.com

Fresh Eggs Daily
by Lisa Steele
160 pages • Hardback • ISBN: 978-0-9855622-5-0

The Herb Lover's Spa Book
by Sue Goetz
192 pages • Hardback • ISBN: 978-0-9892688-6-8

Coffee for Roses
by C.L. Fornari
160 pages • Hardback • ISBN: 978-0-9892688-3-7

Cool Flowers
by Lisa Mason Ziegler
160 pages • Hardback • ISBN: 978-0-9892688-1-3